Great at my job but crap at numbers

Heidi Smith and Mac Bride

For UK order enquiries: please contact Bookpoint Ltd,
130 Milton Park, Abingdon, Oxon OX14 4SB.
Telephone: +44 (0) 1235 827720. Fax: +44 (0) 1235 400454.
Lines are open 09.00–17.00, Monday to Saturday, with a 24-hour message answering service. Details about our titles and how to order are available at www.teachyourself.com

Long renowned as the authoritative source for self-guided learning – with more than 50 million copies sold worldwide – the Teach Yourself series includes over 500 titles in the fields of languages, crafts, hobbies, business, computing and education.

British Library Cataloguing in Publication Data: a catalogue record for this title is available from the British Library.

First published in UK 2010 by Hodder Education, part of Hachette UK, 338 Euston Road, London NW1 3BH.

Typeset by MPS Limited, A Macmillan Company.

Printed in Great Britain for Hodder Education, an Hachette UK Company, 338 Euston Road, London NW1 3BH, by CPI Cox & Wyman, Reading, Berkshire RG1 8EX.

Hachette UK's policy is to use papers that are natural, renewable and recyclable products and made from wood grown in sustainable forests. The logging and manufacturing processes are expected to conform to the environmental regulations of the country of origin.

Impression number 10 9 8 7 6 5 4 3 2 1
Year 2014 2013 2012 2011 2010

Front cover: © kindle_phot
Back cover: © Jakub Semen
Free/Corbis, © agencyby/iSt
iStockphoto.com, © Christo
© zebicho – Fotolia.com, ©
© Photodisc/Getty Images, ©
© Mohamed Saber – Fotolia

Contents

Personal introduction

From Mac:

This book was Heidi's idea. She realized that there are a lot of you out there in essentially non-numerate jobs, with art (or other non-numerate) qualifications, who find yourselves faced with number work. You've been given a budget to manage, or asked to cost a project, or to report on the results of a survey, and it's over ten years since you did Maths for GCSE. You're a well-paid professional in a responsible job – it's a bit embarrassing to ask your boss to remind you how to work out percentages or to do fractions. We're here to help.

When Heidi's other commitments expanded, and she found herself short of time for this project, I was asked to lend a hand, and was very pleased to do so. I used to teach Maths, and I'm very aware that Maths skills fall into the use-it-or-lose-it category. If they don't become a routine part of your life, the techniques and concepts are soon forgotten, though very often it doesn't take much to bring them back into use.

We've made the explanations and examples in this book as simple as possible. That's not because we think you are thick. You're not. You are a very able person and great at your job (it says so on the cover), but we believe that you can only handle number techniques successfully if you understand the principles behind them, and principles are best seen at their simplest.

This book does not try to teach you how to do long division or other complex calculations. There's no point – not in this age of calculators and spreadsheets. They are what you use to do the complex stuff and to get the accurate answers. The problem with them is, however, that if you mis-key, those electronic idiots will give you an answer which looks exactly right, but which will be exactly wrong. What you need, and what we hope you will get from this book, is a sense of numbers and a set of techniques that will enable you to look at the machine's answer and know when it is wrong.

From Heidi:

Many people contributed to the research behind this book and we have gratefully received all the suggestions, ideas and pleading requests for clarification. Most submissions were accompanied by an unequivocal condition of anonymity, so we will not be naming and shaming anyone personally. But we would like to thank everyone who contributed, especially the journalists at *The Guardian* who sent us great ideas for topics, based on their readers' comments.

Finally, this book is for my fiscally-tightened friend 'Andy'. 'Andy' is great at his job.

From Heidi and Mac:

We would both like to thank Alison Frecknall, our highly-numerate commissioning editor at Hodder, for her persistence in seeing through the idea, as well as her vision to adopt the original working title for the final publication.

By the way, the Mac and Heidi characters which appear in some of the examples in this book are entirely fictitious and bear no relation to any person living, dead or anywhere in between.

Available online

As you work through the book you'll find a number of crib sheets and examples marked 'available online'. You can download these from the Teach Yourself website at www.teachyourself.com. Search for *Great at my job but crap at numbers* and follow the links.

You can also download them from a website that Mac has set up for Crap at Numbers stuff. As well as print-ready versions of the time tables and other crib sheets from this book – plus some extra ones – and the example spreadsheets, you will find links to all the websites mentioned in this book – plus some extra ones – and an assortment of other items that he hopes will elucidate and entertain. Visit's Mac's place at: www.crapatnumbers.net.

Only got a minute?

Can maths can make you rich, famous or help you to retire early?

We'd like to say, 'yes, all three', but a more accurate answer would be 'it can certainly improve your chances of two out of three.' Maths does make some people famous – think of Sir Isaac Newton, Rene Descartes, Albert Einstein and Carol Vorderman – but if you were ever going to be a famous mathematician, you would probably be somewhere along that road already, rather than reading this book. (Sorry about the reality-check.)

Being comfortable with and competent in your number work will, however, enable you to make better decisions and to avoid errors at work, and to run your private finances more efficiently.

Whatever it is you do for a living, maths is going to come into it somewhere. It may be a very tiny part of the job – the odd calculation, the end of the year summary of stock or staff – but you need to get it right, both for your own sense of ease and for your career prospects. And whatever the job, once you start on up the career ladder, you will increasingly have to handle numbers. There will be budgets to manage, projects to assess, investment decisions to make, costings to do, and more. None of this requires massive mathematical skills. You do not need to be able to do complex arithmetic in your head or on a sheet of paper. What you do need is to be able to recognize when numbers don't make sense, and to have a clear sight of the bigger picture. Improving your number skills will make it easier for you to move up the career ladder, and to enjoy it more when you get up there. So, that's one way maths can help you to become rich(er).

At home, you will have your own budget to manage, credit cards to balance, and most of you will be buying a house and a car at some point – and typically several of each over the years. These are long-term commitments, where the difference in overall cost of the best and worst deals can run into tens of thousands of pounds or more. If you want to retire early (and happily), you need a good pension and that's another long-term commitment where a poor decision or bad luck can cost a huge amount. The financial crisis of the past few years has hit pensions hard and removed a lot of the old 'certainties' (which only seemed to be certain), so it's a foolhardy writer who will give hard advice on this. All we are aiming to do is give you some of the skills that will enable you to miss the obvious traps and steer towards the better decisions.

Mac's old granny used to say, 'Tak care o' the pennies and the poonds'll tak care o' themselves.' She died broke, apart from the 17 biscuit tins full of coppers under her bed. 'Tak care o' the pennies, and the poonds will leg it while you're no looking' would be closer to the truth. So we offer a maxim for today: 'Take care of the pounds, but keep an eye on the pennies, so they don't dribble away.'

5 Only got five minutes?

Five minutes isn't very long to get you from 'crap at numbers' to 'I can do that', but we'll have a go. First a quick look at some terminology, then we'll show you a couple of tricks.

Grasp the terminology

Happily, there is very little jargon in maths at the level covered in this book, and much of it you may well remember from schooldays. There are a few things that we'd like to pin down now, to avoid any confusion later.

- When you started to do maths at school, in the very early days, you probably called it 'sums' – because that was what you were doing. Sums, or more formally **arithmetic**, is about numbers and techniques for manipulating them. It is one of the bases of **mathematics**, which is about the relationships between quantity, shapes, motion and numbers. This book starts with the sums bit, and moves on from there but concentrates on the practical applications of numbers.
- **Digit**: a single figure in a number. The number '123' has the digits '1', '2' and '3'.
- **Billion**: At one time, 'billion' was taken to mean one million million (1,000,000,000,000) in the UK, and one thousand million (1,000,000,000) in the US. Needless to say, once billions came into everyday use for counting bankers' bonuses and stuff like that, a common meaning had to be agreed, and we fell in line with the US.
- **Decimal place** (d.p.): Way down the other end of the scale, decimal place is a measure of accuracy, and describes the number of digits after the decimal point. If an amount of money is correct to 2 d.p., then it's correct to the pennies.

- **Round numbers:** Approximations, where numbers are rounded up or down to the nearest 10, or 100 or 1000 (or whatever is appropriate). The aim is to get numbers which are easy to handle, so that you can work out the sums in your head or on paper, and get an answer which won't be accurate – rounding has seen to that – but will be the right size. This is sometimes all that is needed to give you an overview of a situation, or it can be used to check the accurate to 7 d.p. figure that the calculator or spreadsheet will give you.

How to check sums

ADDITION

You are given a set of numbers to add up. You can use a calculator for this, but how do you check that you haven't mis-keyed? The answer: check the size and check the end. Here are the numbers:

7,345 + 987 + 4,645 + 12 + 184 + 36 + 8,593

Our calculator result is 21,802. Is this correct?

First check the size. To do this, we convert all those confusing digits to nice round numbers, and as we do this, we will discard the smaller numbers altogether – the answer is going to be in the thousands, so there's no point in worrying about 12 and 36 or even 184.

To get a nice round number, look at the two digits on the left. If the second of these is 5 or more, add 1 to the first digit. Now turn every other digit into a 0. In 7,345 you look at 73, note that 3 is less than 5, and just zero the rest of the digits to give 7,000. In 987, the 8 is more than 5, 9 + 1 makes 10 and we have 1,000.

7,000 + 1,000 + 5,000 + 9,000

Then add the digits: 7 + 1 + 5 + 9 = 22

And tack on the 000s because these are thousands: 22,000

22,000 is close to 21,802. There are no major errors in the calculator value.

Second check the end. Ignore everything except the last digit in each number, and add these.

$5 + 7 + 5 + 2 + 4 + 6 + 3 = 32$

And just keep the last digit – the sum must end in 2. And our calculator result does. As the size is right, and the last digit is right, we can be confident that we have used the calculator correctly.

MULTIPLICATION

We can use similar techniques to check multiplication. Watch. Here's the problem:

476×293

Our calculator has given us: 139,468. Is that right?

First check the size. We'll reduce these to nice round numbers:

500×300

Multiply the first digits: $5 \times 3 = 15$

Then add on a zero for every zero in the two original numbers:

150,000

That's the same size as the calculator's answer. It has the same number of digits and there's not a huge difference between 150 and 139.

Second, multiply the last two digits: $6 \times 3 = 18$

The calculator answer should end with the same digit – and it does.

Practice

Don't believe us? Try it yourself and see how easy it is. Grab your calculator, or run the one in the Accessories on your PC, or fire up the app on your iPhone. For each of these sums, get the gadget to work out the answers then check the size and last digit.

1. $27 + 45 + 8 + 73 + 9 + 62$
2. $346 + 48 + 184 + 5 + 121$
3. $53,196 + 95,274 + 187 + 32,154$
4. 35×21
5. 742×36
6. $7,385 \times 5,783$

Disclaimer

Every effort has been made to make all of the information contained in this book as complete and accurate as possible. However this book should be used only as a general guide and the use of information in this book is at the reader's own risk.

It is sold with the understanding that neither the author nor the publisher are rendering legal, financial, investment or other professional advice or services and any such questions relevant to those areas should be addressed to qualified members of those professions.

1

Basically the basics: simple arithmetic

In this chapter you will learn:
- ***Why and when these skills are useful***
- ***Some mental arithmetic tricks and tips***
- ***How and why to use estimates***
- ***Multiplication the easy way***
- ***How to do division without tears***

Why and when these skills are useful

If you can do simple sums in your head, and not-quite-as-simple ones on a scrap of paper, life will be less stressful and you can be better off. You can save yourself money while you are shopping – whether it's for baked beans, a plane ticket or a mortgage. You can save friendships by turning up on time instead of miscalculating how long it will take to get to the meeting place. You can save your job by spotting the glaring errors in your budget figures before you hand over the spreadsheet to your boss.

And it's not rocket science. Simple arithmetic is more like riding a bicycle – easy once you've got the hang of it. (If you can't ride a bicycle, there's probably a *Teach Yourself* book for that.)

WORKING IN 10s

The counting system is based on 10s. Why? Because we have 10 fingers. (Just as computers count in 2s because they have 2 'fingers' – electric switches can only be on or off.) If you work with the 10s, sums are simpler.

What a nice pair of numbers!

Complements are good. (Not as in 'nice hairdo!' That's a compliment.) The complements of 10 are those pairs of numbers that add up to 10:

1 + 9   2 + 8   3 + 7   4 + 6   5 + 5   6 + 4   7 + 3   8 + 2   9 + 1

You do know these. Just keep them in mind when you are doing addition and subtraction.

Addition

We're going to ignore single digit addition. You can do those. We'll start with numbers of two or more digits. Always add the units, then the tens (and then the hundreds, etc. if there are any). Adding up is easier if one of the numbers ends in 0. Unfortunately, 8 times out of 10 they won't. (8 out of 10? What? Why? See Chapter 7.)

The simplest sums are where the digits add up to less than 10, e.g.:

23 + 45	break the 45 into tens and units, and add the units
23 + 5 = 28	then add the tens
28 + 40 = 68.	

Where the digits will add up to more than 10, break the units in the second number into two bits – one to complement the first, so that you have a nice round multiple of 10. Like this:

27 + 48	split 48 into 3 and 45
27 + 3 + 45	add 3 (the complement of 7) to the first number
30 + 45 = 75.	

An extension of the same trick can be used with bigger numbers:

264 + 387	break 7 into 6 (complement of 4) and 1
264 + 6 = 270 + 381	now turn to the 10s, and split 8(0) into 3(0) and 5(0)
270 + 30 = 300 + 351	you can do the rest yourself

Practice 1

Add:

A 63 + 26
B 36 + 57
C 142 + 53
D 55 + 36 + 113

You will find the answers to the exercises at the end of the chapter.

ADDING IN COLUMNS

If the numbers have two or more digits, and there are more than three numbers, then jot them down in a column, right aligned.

So, if you have got this:

23 + 47 + 117 + 94

Set them out like this:	23 47 117 94	Definitely not like this:	23 47 117 94

Adding single digits down a column is easier than adding bigger numbers to each other, and if it is easier you are more likely to get it right. But you must make sure that the numbers are properly aligned, so that the units are in one column, the tens in one column, etc. Start adding hundred digits to tens digits and you'll be in a real mess.

ADDING BY 10S

What's the best way to add a column of figures? Do you work your way methodically down from the top? No. Look for the digits that add up to 10 and deal with them first. Like this:

23 42 14 77 126 55	23 42 14 77 126 55 Here's one set 3 + 7 = 10 cross them out	23 42 14 77 126 55 And another 4 + 6 = 10 cross them out and that's two lots of 10	23 42 14 77 126 55 Then add what's left in the units 2 + 5 = 7 cross them out that's 2 lots of 10 and 7 = 27
23 42 14 77 126 55 -------- 7 +2 tens	23 42 14 77 126 55 ------------ 7 +2 tens In the tens column: one set 2 + 1 + 7 = 10 cross them out	23 42 14 77 126 55 ---------------- 7 +2 tens We're left with 4 + 2 + 5 = 11 11 + 10 + 2 (tens) we carried over = 23 (tens)	23 42 14 77 126 55 --------------- 37 +2 hundreds Add the hundreds column 1 + 2 carried over Total = 337

Written out like this, it looks a bit long-winded. In practice, it isn't. Try it!

Practice 2

Set out in columns and add using the 10s trick:

A 25 + 68 + 63 + 92 + 47 + 12
B 71 + 215 + 98 + 324 + 62
C 34 + 145 + 232 + 571

Don't sweat the sums

If you have to tackle anything more complicated than the sort of sums that you've just seen here, use a calculator. But to use a calculator properly, you need to be able to do simple arithmetic – you'll see why shortly – which is why this section is here.

Subtraction

Subtraction is addition in reverse, and it's a piece of cake where the digits in the number you are taking away are smaller than those in the first number. For example:

48 – 25

We can do this in two simple steps. First the units:

(4)8 – 5 = (4)3	Do 8 – 5 but keep the 40 in the back of your mind
43 – 20 = 23	Here you are doing 4 – 2 (tens) with the 3 (units) on hold

When there are bigger digits in the second number, you need a different approach. One way is to visualize a number line:

45 – 28

First, do 45 – 8

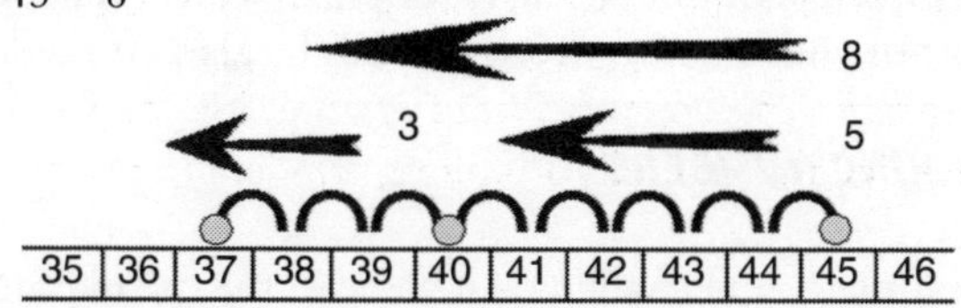

Count 8 steps back along the number line. 5 gets you to 40, with 3 left over, and those take you to 37. The sum is now:

37 – 20

That's just 3 – 2 (tens) and keep the 7 there.

You could also work out 45 – 28 using complements.

First, focus on the units. Break 8 into two parts: what you need to take 45 down to 40 = 5, and what's left over = 3 (8 – 5).

Now it's 40 – (2)3 and the 10s complement of 3 is 7, so 40 – 3 is 37.

Then finish as before: 37 – 20 = 17.

Practice 3

A 67 – 42
B 39 – 28
C 50 – 34
D 72 – 56

Estimates

Estimates are the key to successful and stress-free arithmetic. Why? Chiefly because they help you to avoid errors of scale. If you were selling or buying something, or working out a budget and came up with a total of £15,678.05 or £15,680.75 when it should have been £15,678.50, no one would worry too much – it's a simple slip, a minor difference. But if you came up with £1,567.80 or £156,785.00 that's a whole different ball game. Either way, errors of this magnitude could pose a serious problem – they are the sort that may lose customers, or may lose you your job. Estimates are quick to do; the art of estimating is simple to learn. Make them a routine part of your number work – you'll be glad you did.

NICE ROUND NUMBERS

To do an estimate, you change each value in the calculation to a nice round number. It should only have one, or at most two, significant figures. A significant figure is one in the range 1–9. The rule is to round each value up or down to the nearest 10, or 100 or 1,000 (or 10,000, etc). If the last digit is under 5, round down; if it is 5 or over, round up. Keep in mind which way you round the numbers as it can be useful later.

Some examples:

Value	Nice round number	Keep in mind
26	30	quite a bit under 30
431	400	a fair bit over 400
61,038.23	60,000	not much over 60,000
3.1417	3	not much over 3
8.75	10	quite close to 10

Notice those special technical terms 'quite a bit', 'a fair bit' and 'not much'. The thing is, if the estimates are not much under or over the estimate, then the end result (working on the estimates) won't be much different from the result you would get with the real figures. On the other hand, if you add up a string of numbers and they are all 'quite a bit over' or 'a fair bit over', the end accurate result will be a whole lot over the nice round answer. And it's even more so when you are multiplying (see below).

How much is a fair bit?

You can find the answer to this question, and discover how many tads make a smidgen, at Mac's Crap at Numbers site. Find it at www.crapatnumbers.net

USING ESTIMATES IN ADDITION AND SUBTRACTION

It's worth using estimates even on quite simple sums.
For example:

357 + 64

The nice round numbers from that are:

400 + 60 = 460

The exact answer is 421 – in the same ballpark. The estimated result was a fair chunk higher, but you rounded up 357 by quite a bit, while rounding 64 down by only a little, so you would expect the estimated result to be higher.

Here's a sum that you saw earlier – it was in Practice 1:

71 + 215 + 98 + 324 + 62

We could work out an estimated answer for it like this:

70 + 200 + 100 + 300 + 60

That gives us a rough answer of 730. The accurate answer is 770.

Let's try it with a subtraction:

87 – 28 rounds to 90 – 30 = 60

The exact answer is 59.

Practice 4

For each of these, use estimates to work out the rough answer, then do the sum properly and compare the results.

A 127 + 354 + 417 =
B 28 + 42 + 84 + 221 =
C 86 – 34 =
D 269 – 142 =

Know the difference

Do not confuse the estimates that you do with number work, with the estimates a builder might give you for a job. A number estimate should be of the same order of magnitude (i.e. about the same amount) as the correct answer; a builder's estimate could be a fraction of the actual price for the finished job.

Go forth and multiply

Don't panic! You only need to be able to do simple multiplication – i.e. numbers of two or three digits by values up to 10 – in your head. (And you don't have to do the whole thing in your head – just the multiplying. Use a scrap of paper to record your workings and the result.) For anything else, you can use a calculator. Does this mean that you have to learn (or relearn) your times tables? Yes and no. There are some tricks that give you easy ways of multiplying by most of the numbers up to 10.

To multiply by	Trick
2	Double it.
3	You have to know these, but you should know them already!
4	Double it, then double again.
5	Multiply by 10, then halve the result. $7 \times 5 = 7 \times 10 = 70 \div 2 = 35$.
6	Multiply by 3, then double it – or the other way round
7	Sorry, no trick! You have to learn these. Though there is a way round it, as you will see later.
8	Double it, double again, then again. (It's quicker than it sounds.)
9	Times 3, and again times 3 or use the 'digital calculator' (see below).
10	Add 0 to the end if it's an integer, or move the decimal point one to the right.

Practice 5

A 17×2
B 3×42
C 13×4
D 18×5
E 25×6
F 8×21
G 15×9
H 423×10

Times table: up to 10

1	2	3	4	5	6	7	8	9	10
2	4	6	8	10	12	14	16	18	20
3	6	9	12	15	18	21	24	27	30
4	8	12	16	20	24	28	32	36	40
5	10	15	20	25	30	35	40	45	50
6	12	18	24	30	36	42	48	54	60
7	14	21	28	35	42	49	56	63	70
8	16	24	32	40	48	56	64	72	80
9	18	27	36	45	54	63	72	81	90
10	20	30	40	50	60	70	80	90	100

Times table: the only ones you need to learn

1	2	3	4	5	6	7	8	9	10
2	4	6	8	10	12	14	16	18	20
3	6	9	12	15	18	21	24	27	30
4	8	12	16	20	24	28	32	36	40
5	10	15	20	25	30	35	40	45	50
6	12	18	24	30	36	42	48	54	60
7	14	21	28	35	42	49	56	63	70
8	16	24	32	40	48	56	64	72	80
9	18	27	36	45	54	63	72	81	90
10	20	30	40	50	60	70	80	90	100

Online resource

There are handy print-out-and-keep times tables online. They are available in A4 or pocket/handbag sizes, and go up to 10 or 15.

MULTIPLYING BIGGER NUMBERS

Limiting yourself to the times table may not be much good – by itself – in the real world, but what if you haven't got a calculator to hand? The answer is three-fold.

- Use nice round numbers to get an estimate of the answer – at least you will know more or less how big it will be.
- If you think you can cope with the sum using the grid technique (see below), get out a bit of paper and a pencil.
- If it's too complicated, wait until you can find a calculator – someone will have an app on their iPhone!

Estimating answers

There are three stages to this:

1. Simplify each value to a nice round number, and that means one significant figure followed by zeros. (You can go to two significant figures as long as they are within your times tables, e.g. 110, 1200, 15000.)
2. Multiply the significant figures together and write down the answer.
3. Add up the total number of zeros in those nice round numbers, and write this many after your significant figure answer.

That's it. Let's see it in practice. Last year Laptop Layby Ltd sold 4,678 of their bottom of the range model at £279 each. How much did they take in total? The sum is:

4678×279

In nice round numbers, that is:

5000×300

Calculate:

$5 \times 3 = 15$

Add the zeros:

5000 has three, 300 has two, so the answer will need five
1,500,000

The multiplication grid

Most schools still teach long multiplication, which is a shame because there is a better way to multiply. The problem with long multiplication is that is you have to multiply and carry over the tens, then multiply the next pair and add in the carry, all the way across the line – and if you slip out of alignment, the whole thing goes pear-shaped. The grid solves that, because it separates out the multiplication and the adding, and it keeps things together. It may take a minute longer to set up, but it will save time – and errors – in the end. Watch as we work out the laptop sum.

1 Draw a grid with as many rows as there are digits in one number, and as many columns as there are digits in the other. Draw diagonals through the cells.

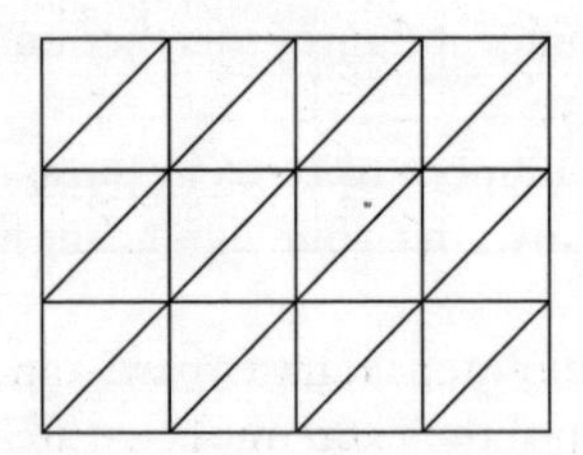

2 Write the first number across the top of the grid and the second down the right hand side.

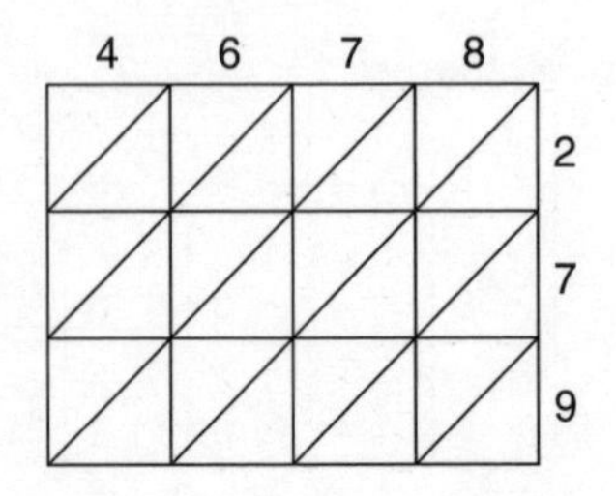

3 Starting anywhere you like – it really doesn't matter – multiply the number at the top of the column with the one at the end of the row and write the answer into the grid, with the 10s value above the diagonal, and the units below.

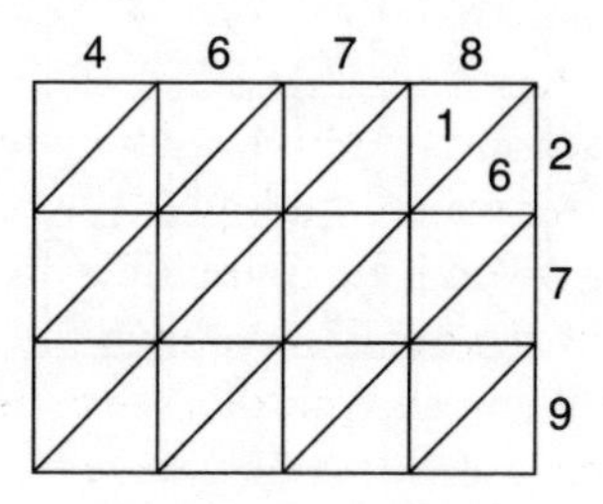

4 Fill in all the cells in the same way – if an answer doesn't have a 10s value, put in a 0. As long as you know your times table up to the 9s, it's a piece of cake.

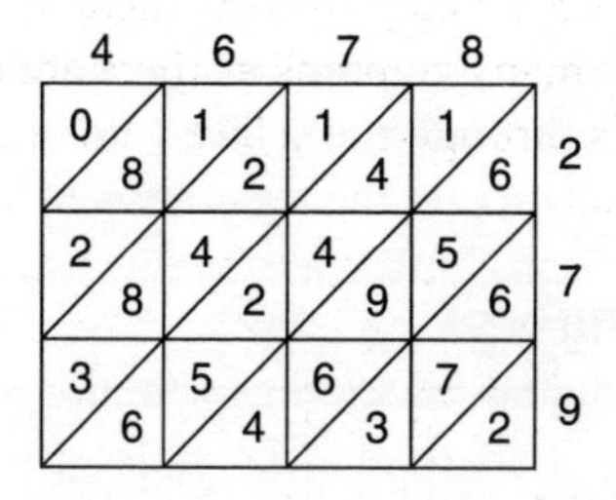

5 Starting from the bottom right corner – and this time it does matter where you start – add up the numbers in each diagonal. Write the units digit at the bottom left of the diagonal. If the total is more than 10, make a note of the 10s digit, and add it into the next diagonal.

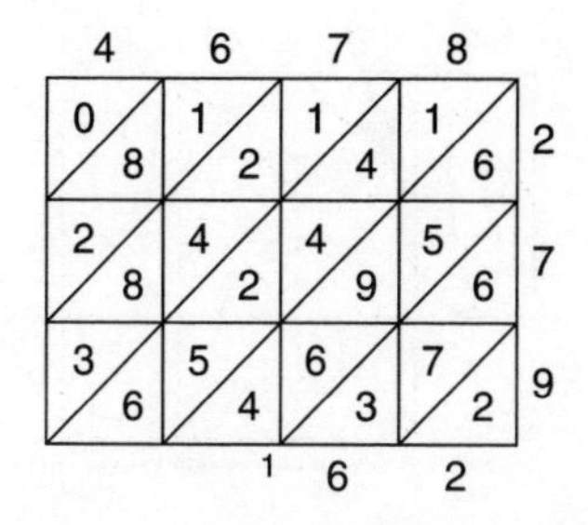

6 Work your way through the diagonals, writing the answers across the bottom and up the left side.

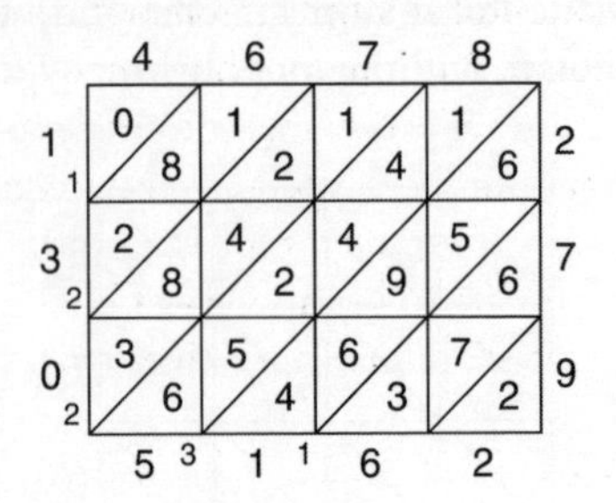

7 Read off the answer from the top left, down and right along the bottom. Here it is 1,305,162.

8 Check against the nice round number estimate. That was 1,500,000 – close enough to give us confidence in the answer.

As we said at the start, the method is a bit fiddlier than long multiplication, but it will normally give you an answer faster, and you are far more likely to get the right answer.

Do it on your fingers!

Here's a special digital calculator for multiplying by 9.

Hold out your hands, fingers spread and imagine them numbered 1 to 10, going left to right. Fold down the finger for the number you want to multiply by 9. The digits on the left give you the 10s value, those to the right of the folded figure are the units. For example, for 4 × 9, fold down digit 4 and you've got 3 tens and 6 units = 36.

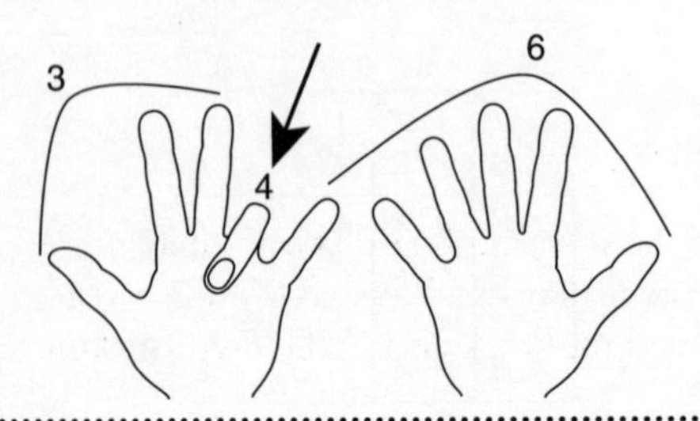

DIVISION

Division is a combination of multiplication and subtraction. When you divide, you are working out how many times you have to multiply one number until it's as big as (or nearly as big as) the other. If you really are crap at numbers, don't even think about trying to work out a division sum accurately if it's in any way complicated. Instead, do a rough estimate to get an idea of how big the answer is, then dig out the calculator and use the estimate to check the result.

Estimating division

There are three stages to this: nice round numbers, knocking off the noughts and multiples.

- First find nice round numbers to replace the amount to be divided (the dividend), and the number you are dividing it by (the divisor). And nice round numbers have one significant figure (between 1 and 9) and the rest are zeros. So, instead of 75 we use 80; instead of 12,345.89 we use 10,000.
- Second, if there is a zero at the end of both numbers, knock it off them both, and keep doing this until one of them runs out of zeros.

$20{,}000 \div 500 = 2000 \div 50 = 200 \div 5$

- Last, run through multiples of the divisor (use your print-out-and-keep times table if necessary) until you get close to the amount to be divided. Don't sweat this. If you've got a really large number divided by a smaller one, stop after the first couple of digits and fill out with zeros. That's your (near enough) answer.

$200 \div 5 = 40$

Let's take a couple of simple examples.

1 A check of the store cupboard shows there are 37 reams of paper. The firm normally gets through 5 reams a day. How long will the stocks last? The division sum is:

$37 \div 5$

Which rounds to

40 ÷ 5

A quick run through the 5 times table gets us to 8 times 5 is 40. We'd rounded up to 40, so the answer is going to be a bit less than 8.

2 Staying with the paper theme, you've got a report to print. Its word count is 19,607 and you can expect to print 420 words per page. How many pages will it take? The sum is:

19,607 ÷ 420

This rounds to

20,000 ÷ 400

Knocking off the noughts gives you

200 ÷ 4

4 times 5 is 20, so 4 times 50 is 200.

Let's just see what happens when you apply the technique to some really nasty numbers – you will be surprised how painless it is. In the last couple of years, the UK government has given the banks £1,353,986,243,876.02 (according to Mac's calculations), and there are 58,543,985 (according to Heidi) people living in the UK. The question is, how much does your bank manager owe you?

The problem:

1,353,986,243,876.02 ÷ 58,543,985

1 Nice round numbers:
1,000,000,000,000 ÷ 60,000,000

2 Knock off the noughts – we can do this three at a time:
1,000,000 ÷ 60 = 100,000 ÷ 6

3. Work through the multiples:
   10 × 6 = 60; 5 × 6 = 30 so 15 × 6 = 90 and that's close enough
   100,000 ÷ 6 = 15,000

Coo, the bank manager owes you £15,000! (Actually, rather more than that – we rounded the amount down and the number of people up.) What do you think the chance is of him lending you £50?

Exactitude corner

For those of you who have to know, the correct answer is £23,127.67*. At something over £15,000 we were in the right ball park. If the calculator had given us £2,312.76 we would have known that we'd keyed in the wrong values. There's more about calculators shortly.

*Except that it isn't correct. We started out with two very dodgy numbers, so no matter how good the calculator is, the result is bound to be equally dodgy, if not more so.

BODMAS, THE RULES OF THE SUMS GAME

If you have a sum that has a mixture of addition, subtraction, multiplication and/or division, then how you work through it can change the outcome. This is not good, so there are rules in arithmetic which tell you the order in which to perform the operations. Computers follow these same rules. The mnemonic is BODMAS, which stands for these operations, and the order in which to carry them out:

1. Brackets
2. Of (to the power of)
3. Division and Multiplication (if there are both, doesn't matter which you do first)
4. Addition and Subtraction (ditto).

Here's a simple for instance. Take this multi-operation sum:

2 + 3 × 4 – 1

If you work through that left to right, you get:

$2 + 3 = 5 \times 4 = 20 - 1 = 19$

Which is the wrong answer.

Here we go again, but following the BODMAS rules:

First do $3 \times 4 = 12$

The sum is now:

$2 + 12 - 1 = 13$

Which is the right answer.

If you need to do things out of order, stick in some brackets. Operations in brackets are performed first. Here are the same numbers again, but with some brackets in place:

$2 + 3 \times (4 - 1)$

First we do $(4 - 1)$. The sum is now:

$2 + 3 \times 3$

Multiplication first ($3 \times 3 = 9$), then the addition: $2 + 9 = 11$

Which is a different answer from last time.

Why do we use the BODMAS order?

When it's just numbers on paper, we use BODMAS because everybody else does, and so that way everyone gets the same answer. Let's see how the BODMAS order works in the real world. Down at the pub the other evening, Mac ordered 2 pints at £3.00, 3 G&Ts at £4.00, a coke at £2.00 and 6 bags of crisps at £1.00 (this pub does not believe in brown change), but being a cheapskate Mac made sure everyone chipped in.

As well as Mac and Heidi, there were 4 of her mates. What did we all pay? This is what the sum looks like in numbers:

$$\frac{(2 \times 3 + 3 \times 4 + 6 \times 1)}{(2 + 4)} \quad \begin{array}{l} \text{work out the total first} = 6 + 12 + 6 = 24 \\ \text{how many people to divide it by?} = 6 \end{array}$$

And the answer is £24 ÷ 6 = £4.00 each.

Practice 6

1. It's Christmas Eve, your plastic's melted, you've got £74.36 in your pockets and 6 presents left to buy. Roughly how much can you spend on each?
2. When 4 of us went on holiday to Wales last summer (it was a walking holiday so we took our wetsuits) we spent £57.83 on petrol. How much – in notes – should we each give the driver?
3. On that holiday, we covered 624 miles. Roughly how much a mile did the petrol cost?
4. Down at the pub, Dodgy Dave is flogging memory sticks for £2.75 each. Well, you know you can sell them on in the office for a fiver each. He's got 17 of them, and you've got £55 in your pocket. Can you afford to buy them all off him?
5. My builder friend, Sasha, paints bathrooms at a rate of 3 m^2 per hour when he's not smoking and at 1.75 m^2 per hour at all other times. Sasha's mate Vova paints at a rate of only 2.5 m^2 per hour, but he can stop Sasha from taking cigarette breaks. Is it better value to employ Sasha and Vova on the project together if I'm in a hurry to get the job done?

SHARING WITH FRIENDS AND ADVANCED ESTIMATING

When are you most likely to face simple division problems in your daily life? For most people, it's when they are splitting a bill with friends, and here near enough is good enough. Round the bill to the nearest multiple of how many are in the party, and work from there.

There are 5 of you in the restaurant, and the bill has come to £137.20. How much do you each have to pay?

If we follow our nice round numbers rule, we'll get to £100 ÷ 5 = £20. While those of you who chip in 20 quid might be happy with this, whoever gets to pay the bill on their card might be less than happy about picking up the £37.20 shortfall on top of their own £20. We need to improve our estimating technique.

We can tackle the problem from the other end. 5 times £20 gives us £100. 5 times £30 gives us £150 – that's closer. It's just over £12 too much, and might even cover the tip. (See Chapter 2.) You want to do the tip separately? Ah well, let's get a bit closer.

£30 each is £12 too much. How about £29 each? That's £5 less in total, but still £7 too much. £28 each gives us £140. That's only £2-ish out overall, less than 50p each and quite near enough. (If your friends are picky about pennies, you need new friends.)

Here's another way to tackle the same problem. We start from our first nice round numbers answer of £20 each.

£20 times 5 covers £100 of the bill, and leaves £37.20 over. We now need to split this 5 ways.

£37.20 ÷ 5 in nice round numbers is £40 ÷ 5 = £8

Add the two results together, and you get £20 + £8 = £28. Near enough (see above).

Answers and explanations

Practice 1

A 63 + 26 = 89
B 36 + 57 = 93
C 142 + 53 = 195
D 55 + 36 + 113 = 204

Practice 2

Set out in columns and add using the 10s trick:

A 25 + 68 + 63 + 92 + 47 + 12 = 307

B 71 + 215 + 98 + 324 + 62 = 770

C 34 + 145 + 232 + 571 = 982

Practice 3

A 67 – 42 = 25

B 39 – 28 = 1

C 50 – 34 = 16

D 72 – 56 = 16

Practice 4

A 127 + 354 + 417 = 898
Estimate 100 + 400 + 400 = 900 (Close!)

B 28 + 42 + 84 + 221 = 375
Estimate 30 + 40 + 80 + 200 = 350 (Close enough)

C 86 – 34 = 52
Estimate 90 – 30 = 60

D 269 – 142 = 127
Estimate (under) 300 – (a lot over) 100 = (a lot less than) 200

Practice 5

A 17 × 2 = 34

B 3 × 42 = 126

C 13 × 4 = 52

D 18 × 5 = 90

E 25 × 6 = 150

F 8 × 21 = 168

G 15 × 9 = 135

H 423 × 10 = 4230

Practice 6

1 £74.36 is around £70. 10 × 6 = 60. 12 × 6 = 72. You can spend a bit over £12 on each person.

2 £57.83 is nearly £60. To split this four ways, we can divide it by two (£30), then by two again (£15). We should each give the driver £15.

3 This is a trick question. It looks as if the sum is 624 miles ÷ £57.83, but that would give you miles per pound, when what you want is pence per mile. We need to turn it round: 5783p ÷ 624 miles. Rounding the numbers gives us 6000 ÷ 600. Knock off the noughts to get 60 ÷ 6 = 10p a mile.

4 £2.75 is a bit less than £3. And 17 is a bit less than 20. £3 times 20 is £60, but the real answer will be quite a bit less than that. £55 will be enough to buy them all – just work out exactly how much they cost before you start negotiating with Dave. (It's £46.75 – offer him £40 for the lot.)

5 The way to approach this is to see how much gets painted in one hour.

Sasha (not smoking)	$3\ m^2$
Sasha (with fag breaks)	$1.75\ m^2$
Vova	$2.5\ m^2$
Sasha + Vova	$3 + 2.5 = 5.5\ m^2$

When Sasha and Vova are there, you will have two wages to pay, so the question is how much will they paint in half an hour, and the answer is $5.5 \div 2 = 2.75\ m^2$. This is more than Sasha paints when he is by himself, so it's worth employing Vova as well. And they'll be out from under your feet in twice the time.

2

Electronic assistance: using a calculator

In this chapter you will learn:

- ***How to use your calculator***
- ***About the buttons***
- ***Why the order of operations matters***
- ***How to use the memory***

How to use your calculator

Calculators – and we're talking about the bog-standard pocket job with buttons for the digits, the arithmetic symbols and a few other bits, not scientific calculators as favoured by engineers, physicists and employees of NASA and Goldman Sachs – are easy to use, and just as easy to misuse. The problem is that they do exactly what you tell them to do, so if you put the wrong question in, you get the wrong answer out – but it will look 'accurate' because it will have umpteen digits after the decimal point. But fear not. You only need to learn four simple skills to become a calculator master.

- ▶ Estimating – so that you can make sure that the calculator result is the right size. (We did this earlier.)
- ▶ What the buttons are for.
- ▶ How to order the sequence of operation to get the right result.
- ▶ Using the memory.

THE BUTTONS

The button sets on calculators – whether they are programs on the computer or on your mobile phone, or actual pocket ones – do vary a bit, but not in any way that matters much to us. You should always find these:

- The digits 0 to 9
- A decimal point
- The arithmetic symbols for Add [+], Subtract [–], Times [×] or [*] and Divide [/] or [÷]
- [=] – equals works out any pending operation and displays the result
- A percent sign [%]
- [Backspace] – deletes the last digit, so you can correct a mis-key (missing from many pocket calculators)
- [CE] – Clear Entry – erases the whole of the last number entered
- [C] – Clear – erases all trace of the current sum
- [M+], [MR], [MC], [M–] and or [MS] – memory keys (some variations here)

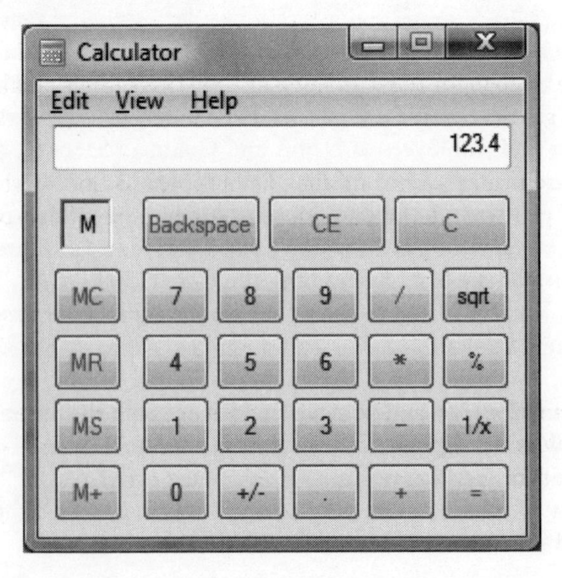

Be clear about the difference between [CE] and [C]. Suppose you had the problem: 12345.6789 × 987654.321, and you keyed in:

12345.6789 [×] 987654.123

Aargh! Got the last 3 digits wrong! If you press [CE] at this point, it will erase the second number, so you can then just finish the sum by keying in:

987654.321 [=]

If you press [C], it will erase the first number as well and set the calculator back to 0.

(Of course, if you have a [Backspace] you can erase and correct just the digits you mis-key.)

The order of operations

Calculators don't know about the BODMAS rules, and they don't normally have brackets, Keep this in mind, because sometimes you will have to change the order in which you do things to make sure that you get the right result.

Most pocket calculators, and the one in Windows and most apps versions, process operations as they are entered. So, if you type in:

2 [+] 3 [×] 5 [=]

You will get 25, because it will work out 2 + 3 = 5, then 5 × 5 = 25.

If you want 2 + (3 × 5), you have to enter the sum the other way round – which is, in fact, the logical way round. Type in first the bit of the sum that you want to be calculated first:

3 [×] 5 [+] 2 [=]

This produces $3 \times 5 = 15 + 2 = 17$.

Unless you are doing a simple string of additions, then, the order in which you enter the sum is equally crucial. For instance, we were in another pub the other evening (hey, everyone has to have a hobby) drinking Desperados. Now, you may call Mac a cheapskate, but this landlord charged for the fruit! £2.99 for a bottle of beer plus 10p for the lime, and that was for 2 of us. How much in total? If you put the sum in the calculator like this:

2 [×] 2.99 [+] 10 [=]

It will give you 15.98 which is clearly way out of whack. What went wrong?

Of course – that 10 means £10. To get 10p you need to put 0.10. Let's try that again:

2 [×] 2.99 [+] 0.10 [=]

That gives us 6.08, which is still the wrong answer. That's the cost of 2 beers and 1 lime wedge. Let's do this thing properly. We need to first work out the total cost of one drink, then multiply: 2.99 [+] 0.10 [×] 2 [=]

Which gives us 6.18. Correct.

Practice 1

Get out your pocket calculator, or find the Calculator app on your iPhone, or start the Calculator program on your PC (it's in the Accessories folder), and try these. Always work out an estimated answer first then compare this with the calculator's answer. If they are not round about the same size, have a careful think about how you are putting the sum into the calculator.

1 Heidi needs a new outfit for this book's launch party. (Launch party? You'll be lucky if you get a lunch out of it, ed.) She's seen a nice frock for £299.99, some shoes at £79.49,

a matching bag at £154.29 and new lipstick at £2.75. How much in total? And how much will be left out of the £500 advance she managed to prise out of the publishers? (Do this as a separate sum at the end.)

2 On that holiday in Wales when we covered 624 miles, we spent £57.83 on petrol. Exactly how much a mile did the petrol cost?

3 We are planning our next book: *Teach Yourself Weekend Holidays in the Sun*, and need some money for research. We need to visit 6 places, on average each flight costs £450, accommodation £200 and expenses £150. And there are 2 of us. How much should we ask the publishers for?

Using the memory

A calculator's memory is a bit like the back of an envelope where you jot down intermediate results when you are working out a longer sum – with two main differences: it can only hold one number at a time, and you can't see what it is while you are working on the next bit of the sum. However, that little bit of memory can make your life easier. Here's how.

STRINGS OF SUMS

The most common use of memory is when you have a mixed string of addition and multiplication – as when totting up bill where there are multiple items of the same type. For instance, what would you charge to drive over and visit a client, given these numbers: it will take 5 hours at £75 an hour, the distance is 48 miles, each way, at 55p a mile, and you'll need lunch (£35 at a nice little Italian place)?

If you type in the sum as a simple sequence, like this:

$5 \times 75 + 2 \times 48 \times 0.55 + 35$

It will give you the wrong answer.

Think how you would tackle the estimated answer – you need to do this anyway to check the calculator's result. You would work out each bit of the bill separately and make a note of the sub-total, then add these together

5 hours @ £75, call it £80. 5 × 8 = 4, so something under £400

2 × 48 miles × 55p, call it 100 miles × 50p = £50

Add those to get £450, then add another £35 for lunch (call it 40) = something around £490.

Using the memory keys, you can work through the sum in the same way that we have just done the estimate – and check each stage against the estimate. The trick is to press [=] after each stage, to get its result, and to use the [M+] to add them into the memory. Watch:

[MC] Clears the memory – not necessary if you've just switched on.

5 [×] 75 [=] Displays 375: you estimated under £400, so that's OK.

[M+] Adds 375 to the memory

2 [×] 48 [×] 0.55 [=] Displays 52.8: almost the same as the estimated £50

[M+] Adds 52.8 to the 375 already in memory.

[MR] This is Memory Recall and it displays the contents of memory: 427.8 in this case

[+] 35 [=] Adds the lunch, and the display now shows 462.80.

STORE FOR LATER USE

The memory is also handy where you have values which you need to use more than once in a calculation. For example, have you come across those websites were prices are given ex-VAT? You pick your bargains, go to the checkout and wham! VAT is added onto the bill and you're having to reach a whole lot deeper into your pocket that you expected. You've selected items costing £34.99, £12.99 and £7.24. What's the VAT and the total including VAT? Here's the sequence to do the job:

[MC] Clears the memory

34.99 [+] 12.99 [+] 7.24 [=] Displays 55.22

[M+] Copies this into memory

[×] 0.175 This is for VAT at 17.5%. We'll be looking at percentages properly in Chapter 5.

[=] Displays 9.6635. The VAT amount (ignore the last couple of digits)

[+] [M+] Add the total from memory

[=] Displays 64.8835 = £64.88

Practice 2

Use the memory buttons to work out the calculations – and remember to estimate the answers so that you can check the results.

1. You've been asked to organize the firm's picnic for 33 people. You've found a caterer who will supply food (and crockery/cutlery) at £8.50 a head. Glass hire is £5.25 a dozen (people can manage with one each). You'll need 4 crates (each of a dozen) of wine at £5.99 a bottle, plus 8 litres of fruit juice at £2.45. What is the total cost?
2. The hospitality fund will cover the first £250 of the cost. How much each will the 33 participants have to pay?

(Contd)

3 Mac returned to his roots last year and went on a haggis-seeking expedition in Scotland. He drove 241 miles the first day and found 12 different haggises (should that be haggae? Hagii?), 268 miles the second day for 7 haggises and 351 miles the third day to locate 3 different haggises. What was his average miles per haggis? Think about this one as you do the estimate, and work out the best way to tackle it on the calculator.

Answers and explanations

Practice 1

1 Heidi's new outfit cost £536.52. Your estimate should have been: £300 (dress), £80 (shoes), £200 (bag) and forget the lipstick – it's so much smaller than the other numbers. Total = £580, and all the big numbers were rounded up so the accurate answer will be less than that. And Heidi has spent more than the book has earned her so far.

2 £57.83 ÷ 624 = £0.0926 which is just over 9p a mile.

3 The sum here is: 6 × (450 + 200 + 150) × 2. The best way to tackle it is to work out the total cost of a trip (450 + 200 + 150) then multiply that. So, 800 × 6 × 2 = £9,600.

Practice 2

1 The sum here is:
(33 × £8.50) + (£5.25 × 3) + (4 × 12 × £5.99) + (8 × £2.45).
The brackets mark the sections that you should add into memory and check with estimates. Here's how they should work out:

Section	Sub total	Estimate
33 × 8.50	280.50	30 × 10 = 300
5.25 × 3	15.75	5 × 3 = 15
4 × 12 × 5.99	287.52	4 × 12 = 48. Call it 50 × 6 = 300
8 × 2.45	19.60	10 × 2 = 20
Total = 603.37		Estimate = 635

2 603.37 – 250 = 353.37. Divided by 33 = £10.71. The estimate here is 300 ÷ 30 = 10.

3 The way to tackle this is to think about the final stage of the sum, which will be total miles divided by total haggises. The last number you will enter will be the haggises, so you need to have this at hand – i.e. in memory – while you are totting up the miles. The process is:

12 + 7 + 3 = 22 [M+]	Haggises in memory
241 + 268 + 351 = 860	Total miles
[÷] [MR] = 39.09	Divide by the haggis number recalled from memory.

3

Unwholesome numbers: fractions and decimals

In this chapter you will learn:

- ***Why and when these skills are useful***
- ***How to visualize fractions***
- ***How to add simple fractions***
- ***Multiplying and dividing fractions***
- ***About working with decimals***
- ***How to convert fractions to decimals***
- ***About multiplying and dividing decimals***

A little bit of jargon

The numbers we are dealing with in this chapter have two parts – the whole number, or *integer*, and the *fraction*. This may be a fraction-as-you-normally-use-the-word, e.g. $\frac{1}{5}$, or a decimal fraction like 0.4. In the explanations that follow, when we say *number* we mean something like $5\frac{1}{3}$, and this is made up of the *integer* 5 and the *fraction* $\frac{1}{3}$.

Why and when these skills are useful

There are two ways of expressing parts of a number: fractions (e.g. 1/2, 3/4, 7/10) and decimals (0.5, 0.625, 0.9). Fractions can be a useful way of describing simple divisions. How do you share 3 pizzas between 5 people? If you put 3 houses on half an acre of land, how big is each plot? The estate is to be divided into 5 parts, with 2 parts bequeathed to the dogs' home, 1 part to the ex-wife and the remaining 2 parts to be shared between the 3 children.

The more complex the fraction, the harder they are to work with, which is where decimals come in. For example, when Mac was a lad (which was *so* last century), fractions were used for most jobs and schoolchildren were expected to be able to do sums like this:

If a book measures $12\frac{5}{8}$ inches by $7\frac{3}{4}$ inches, what is the area of a page?

Give him a piece of paper, a pencil and 10 minutes and he can still work it out, but fractions are not an efficient way to handle partial values like this. (See the next box if you want to know how to do it.) Decimals offer a much simpler approach, and are therefore much more likely to lead to the right answer. That same problem would now read:

If a book measures 12.625 inches by 7.75 inches, what is the area of a page?

Or rather:

If a book measures 32.0675 cm by 19.685 cm, what is the area of a page?

Neither of which are particularly pretty problems, but they will clearly go into a calculator.

Go on then, show us how

First convert the mixed integers and fractions into improper fractions (multiply the whole number by the denominator – the bottom number – then add this to the fractional value):

$$12\tfrac{5}{8} = \frac{12 \times 8 + 5}{8} \quad \text{and} \quad 7\tfrac{3}{4} = \frac{7 \times 4 + 3}{4}$$

(Contd)

Multiply together the top numbers (the numerators), then the bottom numbers:

$$= \frac{101 \times 31}{8 \times 4} = \frac{3131}{32}$$

Now divide the top by the bottom number:

$3131 \div 32 = 97^{26}/_{32}$

Finally simplify the fraction: $97^{13}/_{16}$

Visualizing fractions

Any work with numbers is easier if you have a good mental image of what those numbers mean, and this is especially true of fractions. When you see 2/5, in your mind's eye you should see 2 slices from a cake cut into 5 pieces, or 2 blocks of a 5-block bar of chocolate or 2 bricks in a pile of 5. The image gives you the sense of proportion, of 'about how much' – which gives you the estimate that you need for checking. Here, for example, you should be thinking of 2/5 as a bit under half.

Slices of pie are always good for a mental image – hence the popularity of pie charts – but only for simple fractions:

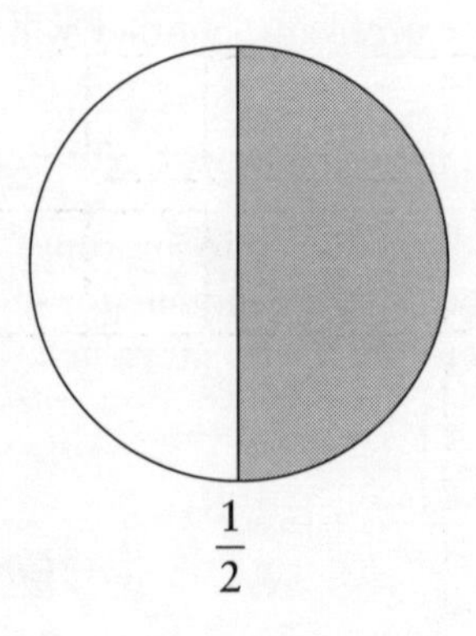
$\frac{1}{2}$

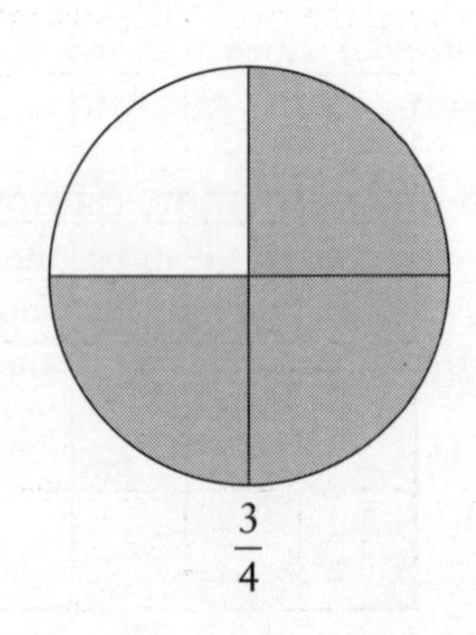
$\frac{3}{4}$

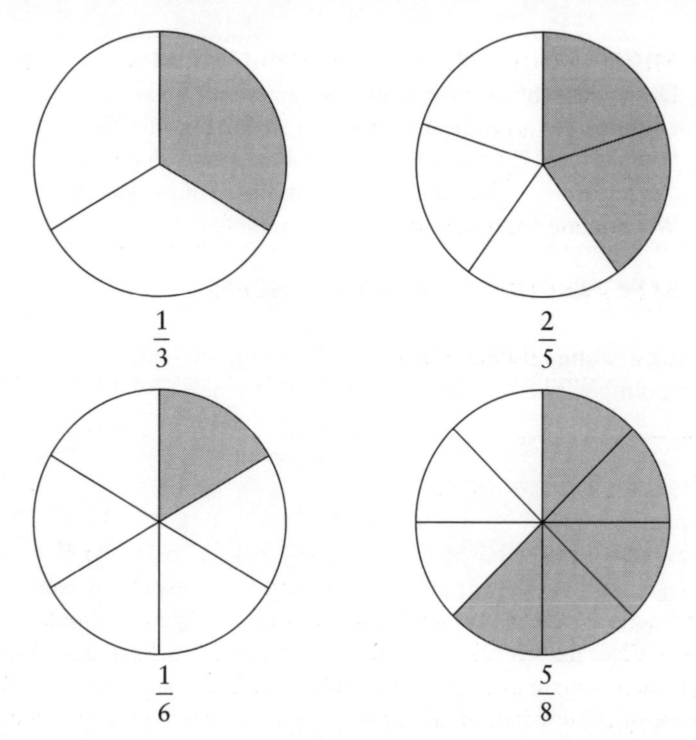

Pies work best with 1/2, 1/3 and 1/4, less well with 1/5 and 1/6, and give poor images for any fractions smaller than 1/8. Blocks and rectangles can handle smaller ones better, but even they have distinct limits. If you can't see at a glance how many blocks there are, then the mental image won't hold – look at the difference between the two 7/10 illustrations.

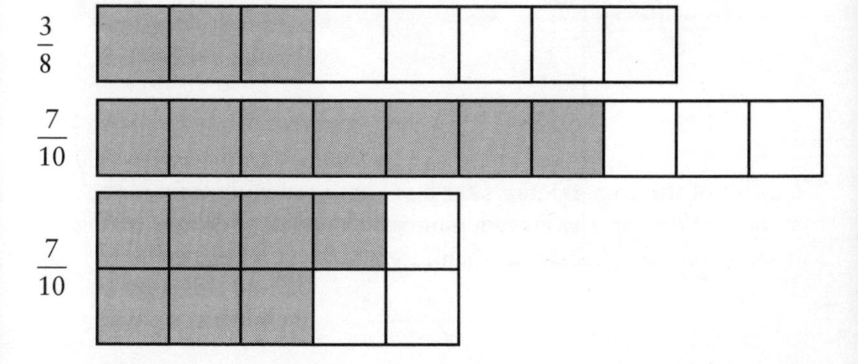

Number 1 rule for working with fractions

Only work with simple fractions – ones that you can visualize. If you come across a sum with a big-number fraction (e.g. 17/43) or a whole-and-fraction value (e.g. $5\frac{3}{8}$), convert the fraction to a decimal, and use your calculator. We'll get on to this shortly.

SAME FRACTION, DIFFERENT NUMBERS

There are often different ways of expressing the same fraction, for example, two quarters are the same as a half (2/4 = 1/2).

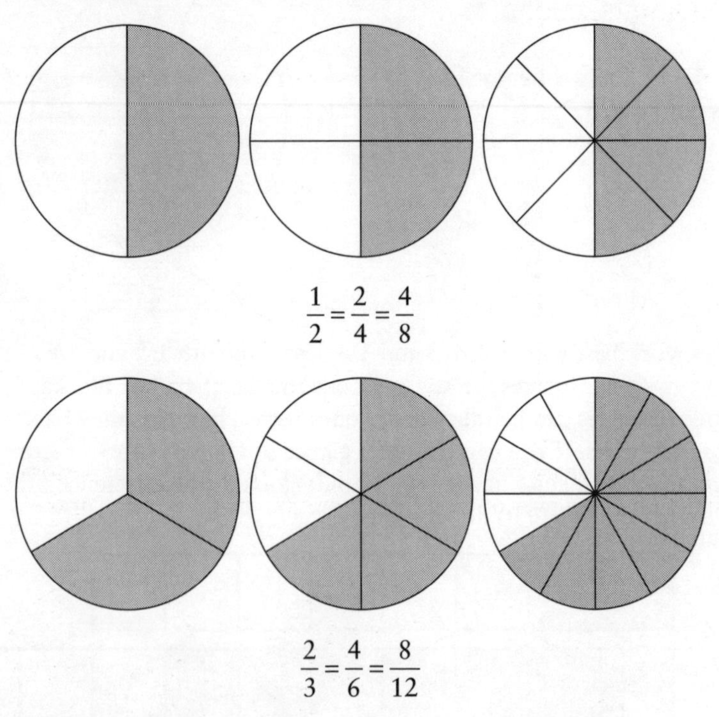

$$\frac{1}{2} = \frac{2}{4} = \frac{4}{8}$$

$$\frac{2}{3} = \frac{4}{6} = \frac{8}{12}$$

Most of the time, fractions are best expressed in their simplest form. If the top and bottom numbers can both be divided by the same number, then divide them, e.g.:

$$\frac{6 \div 2 = 3}{8 \div 2 = 4}$$

Practice 1

Simplify these fractions, by dividing top and bottom by the same numbers – if possible.

1 $\frac{6}{12}$ **3** $\frac{9}{12}$ **5** $\frac{7}{10}$

2 $\frac{8}{10}$ **4** $\frac{12}{16}$ **6** $\frac{15}{20}$

Adding fractions

You can't add different fractions together – not directly. This sum is not do-able:

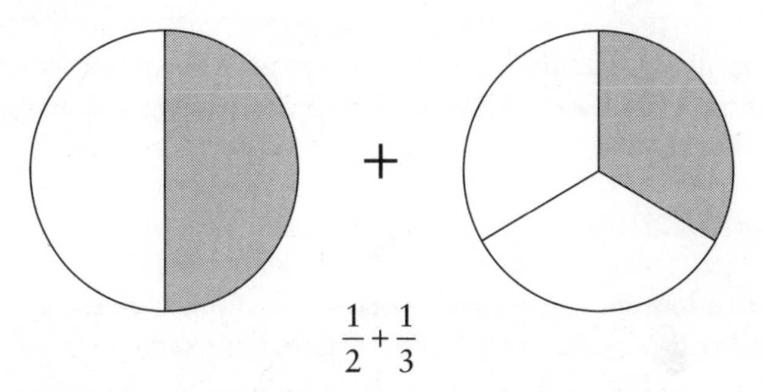

$$\frac{1}{2}+\frac{1}{3}$$

This is because its fractions have different denominators (bottom numbers). However, if we express both fractions in terms of the same denominator, we can add them:

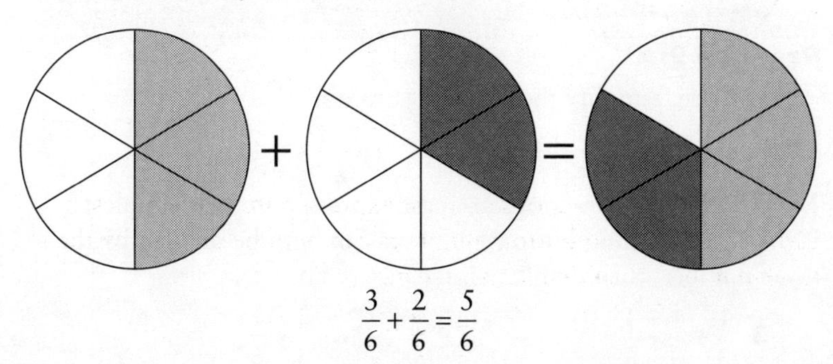

$$\frac{3}{6}+\frac{2}{6}=\frac{5}{6}$$

To find the common denominator, we simply multiply the denominators. For example, to add quarters and thirds, we would have to work in twelfths, because $4 \times 3 = 12$.

To convert two fractions before adding, we multiply both parts of each fraction by the denominator of the other.

$$\frac{1}{4}+\frac{1}{3}=\frac{1\times 3}{4\times 3}+\frac{1\times 4}{3\times 4}=\frac{3}{12}+\frac{4}{12}=\frac{7}{12}$$

If the resulting fraction can be simplified, do so. $\frac{7}{12}$ is as simple as it gets.

$$\frac{1}{2}+\frac{2}{3}=\frac{1\times 3}{2\times 3}+\frac{2\times 2}{3\times 2}=\frac{3}{6}+\frac{4}{6}=\frac{7}{6}$$

If the resulting fraction is top-heavy, convert it to an integer and a fraction. Here we can take 6/6 (= 1) away, leaving 1/6, and giving us the answer $1\frac{1}{6}$.

SUBTRACTION

As with addition, you can only subtract fractions with the same denominator, so convert first if necessary. For example,

$$\frac{1}{2}-\frac{1}{3}=\frac{1\times 3}{2\times 3}-\frac{1\times 2}{3\times 2}=\frac{3}{6}-\frac{2}{6}=\frac{1}{6}$$

Practice 2

Try these. Simplify the result if you can.

1 $\frac{1}{2}+\frac{1}{4}$

2 $\frac{1}{4}+\frac{2}{3}$

3 $\frac{1}{3}+\frac{1}{3}+\frac{1}{6}$

4 $\frac{3}{4}+\frac{1}{2}$

5 $\frac{3}{5}+\frac{1}{2}$

6 $\frac{2}{3}+\frac{1}{6}$

There's a limit!

Adding and subtracting fractions is not something that you are likely to want to do very often, and you should never have to do this with anything more complicated than halves, quarters, thirds, sixths and the like.

Multiplying by fractions

These are the most common sorts of sums involving fractions that most of us see, and they are normally about sharing costs or other things. You know you've got a multiply by a fraction job when it says 'of' in the middle:

1/4 of 18, 2/5 of 30, 3/4 of 240.

Written as sums these are:

$1/4 \times 18$, $2/5 \times 30$, $3/4 \times 240$.

And they are easy. Multiply by the top bit of the fraction, and divide by the bottom bit – though it's often easier to do the division first. So:

$1/4 \times 18 = 18 \div 4 = 4\frac{1}{2}$
$2/5 \times 30 = 30 \div 5 = 6 \times 2 = 12$
$3/4 \times 240 = 240 \div 4 = 60 \times 3 = 180$

Most multiplication sums will involve a fraction and an integer. Some will have fractions on both sides, e.g. a half of a quarter (1/2 of 1/4). With these, multiply the top numbers together, then multiply the bottom numbers together. The result may need simplifying.

$$\frac{1}{2} \times \frac{1}{4} = \frac{1 \times 1}{2 \times 4} = \frac{1}{8}$$

$$\frac{2}{5} \times \frac{3}{4} = \frac{2 \times 3}{5 \times 4} = \frac{6}{20} = \frac{3}{10}$$

Piece of cake – as long as you only do it with simple fractions. (Remember Rule Number 1.)

DIVIDING

In the real world there's not a lot of call for dividing things by fractions – and when there is, the problem can usually be dealt with another way. For instance, 'If I buy 7 pizzas and divide them into quarters, how many will that feed?' can be expressed as this sum:

$7 \div 1/4$

But if you pose the question the other way: 'If I buy 7 pizzas and cut each into 4, how many will that feed?' you get this sum:

7×4

which is a much nicer one. (Though the answer is still zero – a quarter of a pizza won't feed anybody!)

However, if you do ever find yourself faced with dividing by fraction sums, there's a simple way to deal with them. Turn the fraction upside down and multiply. (Which is what we did with the pizza problem.) Instead of:

$24 \div 3/5$

write it as:

$24 \times 5/3$

and work it out:

$24 \times 5 = 120 \div 3 = 40$

As a rough check, remember that if you are dividing by a value less than 1, you will get more, and the smaller the fraction, the bigger the end result will be. As you saw earlier, dividing by 1/4 is the same as multiplying by 4. If you divided the pizzas into 1/6, you'd get 6 slices per pizza.

Practice 3

Try these. Simplify the result if you can.

1 2/3 × 15
2 3/4 × 32
3 27 × 1/3
4 1/3 × 1/2
5 1/4 × 2/5
6 3/4 × 2/3

Practice 4

Each of these problems can be solved by a sum involving fractions. Express them as sums and find the answers.

1 It's half a mile from Heidi's place to the gates of the Park, three quarters of a mile across to the gates on the opposite side, five eights of a mile from there down to the canal, and half a mile back along the canal to Heidi's place. How far is her morning jog?

2 When Uncle Ebeneezer died, he left £150,000 to be divided this way: one third to his faithful housekeeper of 6 months, Fifi LaTouche, one sixth to his only son, and the remainder to be divided equally between his five daughters. How much did each daughter get? What fraction was this?

3 At the village bring-and-buy sale, Marge and Flora were both selling cake. The cakes were the same size and thickness. Marge cut hers into 6, then cut each of these pieces into 4. Flora cut hers in half, then cut each half into 5. Marge was offering 2 pieces for 50p; Flora offered 1 piece for 50p. Which was best value?

Decimals

Once you get away from simple single-digit fractions, it is normally easier and clearer to express part values in decimals – and decimal fractions have the added bonus that they can be handled by a calculator.

Now, this next bit may be very obvious, but there is a real point to it, so bear with us.

Decimal fractions are an extension of the normal number system, where the value of a digit within a number depends upon its place. As you go left from the decimal point, each digit is worth 10 times more, and as you go to the right each digit is worth 10 times less. So, 123.45 is worth:

1	2	3	4	5
×	×	×	×	×
100	10	1	1/10	1/100

What this means is that 0s can be crucial for establishing value. We are all of us pretty good at remembering them with integers, but may not be so hot about remembering them on the other side of the decimal point. There's a lot of difference between 0.5 (1/2) and 0.05 (1/20) or 0.005 (1/200) if you are using that fraction to calculate shares.

How many decimal places?

Some multiplication and division sums can produce long strings of digits after the decimal point. In many cases, many of these are not necessary. If you are dealing in money, you would normally only bother with two digits after the point – representing the pence or cents. If you are a NASA scientist calculating the trajectory of a space flight, you would want it accurate to 20 or more decimal places, otherwise at the end of your craft's 56 billion mile journey, it might be 20 miles out and totally miss its target.

ADDITION AND SUBTRACTION

The key thing here is that you must line up the numbers on the dots. Apart from that, addition and subtraction involving decimals is exactly the same as with integers.

123.45 + 56.789 + 32.1 + 0.007

Line them up!

```
  123.45
   56.789
   32.1
    0.007
= 212.346
```

Quick check: 120 + 60 + 30 = 210. And that's near enough.

Checking is essential. One of the most common calculator errors is missing out the decimal point. We all mis-key occasionally, but you are much less likely to notice if you fail to connect with the dot key, than with a number key – the dot may be just a single pixel on a pocket calculator. The rule when checking sums involving decimals: ignore the decimals unless they are more or less all you've got. So, for instance, if the sums involve money and most of the amounts are pounds and pence, ignore the pence, but if most of the sums are pence then round them to the nearest 10p.

Practice 5

Use your calculator to find the answers to these sums. In each case, check the answer with a nice round number estimate.

1. 555.34 + 12.895 + 107.107 + 8765.76 + 0.002
2. 0.125 + 0.0125 + 0.000125
3. 1.99 + 4.99 + 12.49 + 0.25 + 0.33
4. 1.99 + 0.99 + 0.49 + 0.25 + 0.33
5. 1000 – 234.56
6. 987.24 – 76.127

Decimals make life so much easier

In the dim and distant past, before D-Day – that's D for Decimalisation – when Mac was still a lad, people in the UK measured things in yards, feet and inches, weighed things in stones, pounds and ounces, and paid for it all in pounds, shillings and pence. So you would have problems like this:

(Contd)

What is the cost of 2 stone, 5 pounds and 7 ounces of bird seed if it costs 3 pounds two shillings and fourpence halfpenny a pound? (There are 16 ounces to a pound and 14 pounds to a stone; 12 pence to a shilling and 20 shillings to the pound.)

Primary school children did sums like these. Mind, it could take so long that by the time you had finished, the local pigeons would have eaten 10 ounces and the price gone up by sixpence a pound.

So, let's raise a 125 ml glass of French wine to Napoleon Bonaparte who started the whole decimalisation ball rolling.

MULTIPLYING DECIMALS

One of the things to remember about multiplying decimals is that they are fractions, and if you multiply a decimal (fraction) by another decimal (fraction) the answer will be smaller than either of them. And this is just as it is with simple fractions – a half of a half is a quarter. No matter how big the numbers, if they have digits after the decimal point, then the answer will have a longer string of digits after the decimal point. And there is a simple way to know how long that string will be, which gives you a way to find and/or check the answer.

The number of decimal places (d.p.) in the result will be the sum of the number of decimal places in each of the values being multiplied.

12.3 (1 d.p.) × 3.12 (2 d.p.) = 36.376 (3 d.p.)

Unfortunately you cannot rely on this to check a calculator's result. If one of the numbers ends in 5, and the last digit of the other number is even, the result will end in 0, and the calculator won't display it. For example:

0.4 (1 d.p.) × 0.5 (1 d.p.) = 0.20 (2 d.p.)

But the calculator will show: 0.2.

We need a better way of checking. If in both numbers, the decimal fraction is just the tail end of a larger number, then – for checking purposes – ignore the decimals and treat them as integers. (See Chapter 1, page 11.)

If one of the numbers is less than 1, here's what you need to do to get an estimate to check the result.

1 Find the significant digit (1 to 9), but make sure that you keep any leading zeros between it and the decimal point. Some examples:

0.0813	becomes	0.08	
0.0004687	becomes	0.0005	(4687 rounds to 5000 and we drop the 000)

2 Count how many steps it would take to move the decimal point so that it is immediately to the right of the significant digit. To get from 0.03 to 3.0 takes 2 steps; 0.005 to 5.0 takes 3 steps
3 Round the other number and multiply by the significant digit: e.g. in 1200 × 0.005, calculate 1200 × 5 = 6000.
4 Move the decimal point back as many steps as you just moved it forward.

Here are two examples. In the first we have a large number and a decimal fraction:

1

The sum:	1234.56 × 0.6789
Rounds to:	1200 × 0.7
Take the steps:	0.7 to 7.0 is one step
Do the sum:	1200 × 7 = 8400
Step back:	8400 step back one = 840
Calculator result:	838.14278

Really rough check: 0.67... is a bit more than half. Half of 1200 is 600, so the answer should be a bit more than 600.

In the second example, we've got two decimals, both less than 1.

2 The sum: 0.1234×0.056

Rounds to: 0.1×0.06

Take the steps: 0.1 to 1.0 is one step; 0.06 to 6.0 is two steps – three steps in all.

Significant digits: $1 \times 6 = 6$

Step back: 6.0 step back three = 0.006

Calculator result: 0.0069104

Really rough check: both numbers were small, and a small bit of a small bit is going to be tiny.

Practice 6

This method looks a bit fiddly, but there is nothing difficult about it. Once you have used it a few times, you will see how easy it really is. Have a go with a few sums now. Work out the estimate, then use your calculator to find the exact answers. Does your estimated answer have the decimal point in the same place? Does it start with the same number (or one very close)?

1 12.95×0.66
2 153.33×0.0892
3 55532.19×0.0022
4 0.5×0.25
5 0.8×0.001

Dividing by decimals

Division is pretty much the mirror image of multiplication, so the same techniques apply – with minor differences. We can summarize:

1 Use a calculator to get the accurate answer. Use your head (and perhaps a bit of paper) to get a rough estimate as a check.
2 If both numbers are more than 1, ignore the decimal fraction and treat at integers. (See Chapter 1, page 15.)
3 Remember that when you divide by a decimal fraction (or by any fraction) the answer is going to be bigger than what you started with. At the simplest: 1 divide by 0.5 is the same as saying how many times will 0.5 go into 1, and the answer is 2.

4. Simplify the decimal to a nice round number, and count how many steps it would take to move the decimal point to its right.
5. Do the division, as if they were both integers.
6. Add a zero for every step that you took in stage 4.

Example:

1234.56 ÷ 0.378

Rounds to: 1200 ÷ 0.4
Take the steps: 0.4 to 4.0 takes one step
Do the division: 1200 ÷ 4, 4 × 3 = 12, so 4 × 300 = 1200
Add the zeros: 00 + one step = 3000
Calculator result: 3266.0317

Practice 7

Use your calculator to find the answers to these sums. In each case, check the answer with a nice round number estimate.

1. What's the cost of 43.57 litres of petrol at 115.35p a litre? Answer in £s, please.
2. You've spend 11 hours 45 minutes on a job for a client, and your time is charged at £75.80 an hour. What will the cost of your time?

Converting fractions to decimal

We kept saying, 'Don't sweat over complicated fractions – turn them into decimals.' But how? Easy! Use your calculator and divide the top number by the bottom. So,

2/5 = 2 ÷ 5 = 0.4

3/8 = 3 ÷ 8 = 0.375

17/43 = 17 ÷ 43 = 0.3953488 (and aren't you glad to have a calculator)

But what if it's a mixed number – with a whole and fraction? No problem. Just feed it into the calculator in the order: top ÷ bottom + whole. (And you must do it that way round.) If necessary, store that decimalized number in memory while you convert another fractional value.

$7\frac{3}{5} = 3 \div 5 = 0.6 + 7 = 7.6$

Never ending decimals

Note that some conversions – as with any division sum – can produce never ending decimal strings. 1/3 as a decimal is 0.33333333333333333333... 1/6 is 0.16666666666666666.... 1/7 is 0.142857142857... With anything like this, make a decision about how accurate you need to be and round the decimal places off at that point. For money sums, 2 places is usually what is needed; for real world weights and measure, 3 decimal places is often enough.

Practice 8

Convert these fractions to decimals.

1 $\frac{1}{4}$

2 $\frac{2}{3}$

3 $\frac{4}{9}$

4 $\frac{7}{10}$

5 $\frac{5}{6}$

6 $2\frac{3}{4}$

7 $45\frac{4}{5}$

8 $3\frac{1}{7}$

9 $47\frac{11}{12}$

10 $\frac{5}{8}$

Ready-reference: Fraction–decimal equivalents

Here's a set of fractions and their decimal equivalents which you may find useful. The decimals are given to 3 places of accuracy.

$\frac{1}{2}$	= 0.5
$\frac{1}{3}$	= 0.333
$\frac{1}{4}$	= 0.25
$\frac{1}{5}$	= 0.2
$\frac{1}{6}$	= 0.167
$\frac{1}{7}$	= 0.143
$\frac{1}{8}$	= 0.125
$\frac{1}{9}$	= 0.111
$\frac{1}{10}$	= 0.1

There is a print-ready version of this and a fuller table available online.

So why isn't it all decimals?

Nearly 40 years after decimalization in the UK, we still use lots of old Imperial measures – ask anyone their height and they will answer in feet and inches; we talk about miles per gallon but petrol is sold by the litre; we drink beer by the pint; and – this one's great – we describe the length of timber in metres and its other measurements in inches, so you get '2.4 metres of 4 by 2'.

The UK is not alone in being slow to accept new systems. In France you can buy produce at the markets in *livres* (about the same as UK pounds), 200 years after the introduction of kilograms. They also give house prices in francs, 10 years after the switch to the euro, and some people still use the old franc for house prices. (The franc was devalued in 1960 at the rate 100 old francs = 1 new franc.)

Answers and explanations

Practice 1

1 $\frac{6}{12} = \frac{1}{2}$

2 $\frac{8}{10} = \frac{4}{5}$

3 $\frac{9}{12} = \frac{3}{4}$

4 $\frac{12}{16} = \frac{3}{4}$

5 $\frac{7}{10} = \frac{7}{10}$

6 $\frac{15}{20} = \frac{3}{4}$

Practice 2

1 $\frac{1}{2} + \frac{1}{4} = \frac{2}{4} + \frac{1}{4} = \frac{3}{4}$

2 $\frac{1}{4} + \frac{2}{3} = \frac{3}{12} + \frac{8}{12} = \frac{11}{12}$

(4 × 3 = 12 to get the common denominator)

3 $\frac{1}{2} + \frac{1}{3} + \frac{1}{6} = \frac{3}{6} + \frac{2}{6} + \frac{1}{6} = \frac{6}{6} = 1$

4 $\frac{3}{4} + \frac{1}{2} = \frac{3}{4} + \frac{2}{4} = \frac{5}{4} = 1\frac{1}{4}$

5 $\frac{3}{5} + \frac{1}{2} = \frac{6}{10} + \frac{5}{10} = \frac{11}{10} = 1\frac{1}{10}$

6 $\frac{2}{3} - \frac{1}{6} = \frac{4}{6} - \frac{1}{6} = \frac{3}{6} = \frac{1}{2}$

Practice 3

1 2/3 × 15 = 2 × 15 ÷ 3 = 30 ÷ 3 = 10

2 3/4 × 32 = 3 × 32 ÷ 4 = 96 ÷ 4 = 24 (You could have done the division first, i.e. 32 ÷ 4 = 8 then 3 × 8 = 24.)

3 27 × 1/3 = 27 ÷ 3 = 9

4 1/3 × 1/2 = 1 × 1 ÷ 3 × 2 = 1 ÷ 6 = 1/6

5 1/4 × 2/5 = 1 × 2 ÷ 4 × 5 = 2 ÷ 20 = 2/20 = 1/10

6 3/4 × 2/3 = 3 × 2 ÷ 4 × 3 = 6 ÷ 12 = 1/2

Practice 4

1. The sum of Heidi's run is: 1/2 + 3/4 + 5/8 + 1/2
   We can simplify this to: 1/2 + 1/2 = 1 + 3/4 + 5/8
   Common denominator is 8: 1 + 6/8 + 5/8 = 1 + 11/8 =
   1 + 1 + 3/8 = 2⅜
2. There are several sums here:
   1/3 of £150,000 = £50,000 for Fifi LaTouche
   1/6 of £150,000 = £25,000 for the son
   That leaves: £150,000 - £50,000 - £25,000 = £75,000
   As a fraction: 1/3 + 1/6 = 2/6 + 1/6 = 3/6 = 1/2 (to Fifi and the son)
   1 – 1/2 = 1/2 the fraction left
   £75,000 ÷ 5 = £15,000 for each daughter
   As a fraction: 1/2 ÷ 5 = 1/10
3. Marge cut her cake into 6 × 4 = 24 pieces, but sold them 2 at a time. 2/24 = 1/12.
   Flora cut hers into 2 × 5 = 10
   1/10 is more than 1/12. Flora's cake was the best value.

Practice 5

1. 555.34 + 12.895 + 107.107 + 8765.76 + 0.002 = 9441.104
   500 + 10 = 610 + 110 = 720 (call it 700) + 9000 =
   9700 + 0 = 9700 Check!
2. 0.125 + 0.0125 + 0.000125 = 0.137625
   0.1 + 0.01 = 0.11 + 0.001 = a bit more than 0.11 Check!
3. 1.99 + 4.99 + 12.49 + 0.25 + 0.33 = 20.05
   2 + 5 = 7 + 12 = 19 and a bit more Check!
4. 1.99 + 0.99 + 0.49 + 0.25 + 0.33 = 4.05
   2 + 1 = 3 + 0.5 = 3.5 + 0.3 = 3.8 + 0.3 = 4.1 Check!
5. 1000 – 234.56 = 765.44
   1000 – (a bit more than) 200 = (a bit less than) 800 Check!
6. 987.24 – 76.127 = 911.113
   1000 – 100 = 900

Practice 6

1. 12.95 × 0.66 = 13 × 7 = 13 × 7 (1 step) = 91, step back to 9.1
   (Calculator result: 8.547)

2. $153.33 \times 0.0892 = 150 \times 0.09 = 150 \times 9$ (2 steps) = 1350, step back to 13.5 (Calculator result: 13.677036)
3. $55532.19 \times 0.0022 = 60000 \times 0.002 = 60000 \times 2$ (3 steps) = 120000, step back to 120 (Calculator result: 122.170818)
4. $0.5 \times 0.25 = 0.5 \times 0.3 = 5 \times 3$ (2 steps) = 15, step back to 0.15 (Calculator result: 0.125)
5. $0.8 \times 0.001 = 8 \times 1$ (4 steps) = 8, step back to 0.0008 (Calculator result: 0.0008)

Practice 7

1. $43.57 \times 115.35 = 5025.7995$p = £50.26. The estimate would be a bit over 40 × a bit over £1, giving an answer of a fair chunk over £40, which is what we have.
2. 11.75×75.80 = £890.65. The estimate is a bit over 10 × a chunk under £80 giving something around £800, which we have.

Practice 8

1. 1/4 = 0.25
2. 2/3 = 0.6666667
3. 5/6 = 0.8333333
4. 7/10 = 0.7
5. 4/9 = 0.4444444
6. $2\frac{3}{4}$ = 2.75
7. $45\frac{4}{5}$ = 45.8
8. $3\frac{1}{7}$ = 3.1428571
9. $47\frac{11}{12}$ = 47.91666667
10. $7\frac{5}{8}$ = 7.625

4

Doing business over there: common conversions

In this chapter you will learn:

- ***Why and when these skills are useful***
- ***Simple ways to convert between different systems of weights and measure***
- ***About Fahrenheit and Celsius scales***
- ***How to handle currency conversion***
- ***How not to worry about currency conversions***
- ***How to handle time zones***

Why and when these skills are useful

Oh life would be much simpler if there was only one system of weights and measures, and a common world currency, but there isn't. We don't even have a single system of weights and measures in the UK, and use the old imperial and the metric measures willy-nilly. We sell water by the litre, beer by the pint, wine by the 75cl bottle and petrol by the gallon! We measure distances in miles, materials in metres and ourselves in feet and inches. And that's even before getting on the boat or plane that will take you 'over there' to where they do things differently again. But we muddle by with our mixed measures in our daily life, so we are already well on the way to handling conversions over there.

The secret of stress-free conversion

Most conversions are done in basically the same way, because you are doing basically the same thing – changing the units that you use to describe a quantity. The quantity may be length, weight or volume and the units of measurement may be different, but the process is the same.

There are four sides to stress-free conversion:

1. **The rough check.** You need to know which unit of measurement is bigger, and roughly by how much. For instance, 1 litre is a bit less than 2 pints. If you keep this in mind, you can do a rough check on the answer. If you started with litres, you should have a bit less than twice as many pints. If you started with pints, you should have a bit more than half as many litres. And this tells you what sort of sum you have to do. To go from litres to pints, you need to multiply – to get a bigger number. To go from pints to litres – and get a smaller number – you need to divide.
2. **The magic number.** This is the value that you use to convert one unit into another. To turn a volume in litres into pints, you multiply by 1.759,753,986. No, just kidding! That figure is for rocket scientists only, and way too accurate for any normal use. 1.75 will do very nicely. And if you are converting the other way – pints into litres, you divide by the same magic number.
3. **The benchmarks.** It helps if you remember a few benchmark values – ones that you can visualize. A pint (of beer) is a half litre (when you're over there). 8 pints (or 1 gallon) is 4.5 litres – call it 5 if you like – so if you can get 50 miles to the gallon out of your car, a 200 mile journey will take about 4 gallons, or 20 litres.
4. **The crib sheet.** There's a handy print-it-and-pocket-it sheet of magic numbers on the Crap At Numbers website.

The fly in the ointment

You may have spotted the fly in the ointment of our stress-free conversion secret – though we've been talking about *units of measurement*, the problem is that there are *systems of units of measurement*. Sometimes it's not enough to know the magic number to convert miles to kilometres. Sometimes you will start with mixed units – feet and inches, or stones and pounds – and there's no single magic number here. But fear not! For situations like this, you use the Mark II, two-stage stress-free conversion technique.

1 Convert the mixed units to one unit, e.g. feet and inches to all inches – and there are magic numbers for this.
2 Convert the quantity in the single unit to the other system, e.g. inches to centimetres.

We'll deal with mixed jobs as we go through.

Going the distance

Let's start from the top and work our way down to smaller numbers.

MILES AND KILOMETRES

When might you need to convert between these? The obvious answer is when planning a trip in Europe (or elsewhere), where distances are measured in kilometres but you are more comfortable thinking of them in miles. Another common situation would be when you needed to compare distances given in miles with those in kilometres.

The rough check

Miles are bigger than kilometres. One mile is a bit more than one and a half kilometres. If you are not happy with halves and will take a really rough check, think of it as quite a lot less than twice as many. Going the other way, a mile is a bit more than half a kilometre.

The magic number

1 mile = 1.61 kilometres, so the magic number is 1.61.

You can also think of the miles to kilometres number as 8/5 (multiply by 8 then divide by 5, or divide by 5 and multiply by 8 – whichever is easier). With kilometres to miles, the fraction is the other way up 5/8 (a bit more than half, remember).

With these numbers you may be able to do sufficiently accurate conversions in your head or on a scrap of paper.

Benchmarks

100 m = 160 km (well nearly)
100 km = 60 m (and a bit over)

Examples

It's 1149 miles from London to Rome. What's that in kilometres?
Rough check: a bit more than one and a half times 1100 miles is about 1600 km.
Magic number: (on the calculator) $1149 \times 1.61 = 1849.89$
(using the fraction, on paper) $1150 \div 5 = 230 \times 8 = 1840$

It's 1444 km from Rome to Paris. What's that in miles?
Rough check: a bit more than half of 1400 is something over 700 miles.
Magic number: (on the calculator) $1444 \div 1.61 = 896.89$
(using the fraction, on paper) $1440 \div 8 = 180 \times 5 = 900$

Practice 1

1. Heidi and Mac are going to Paris (it's a research trip, so tax-deductable). Heidi's car can go 240 miles on one tank full, but it will be half empty by the time it gets to Le Havre. Can we drive the 196 km from there to Paris without stopping?
2. It's 3.3 km from the Place de la Concorde (Paris) to the Jardin du Luxembourg. Mac can't walk more than 2 miles without stopping for a beer. Can he get there in one go?
3. The circumference of the Earth is 24,901 miles. What's that in kilometres?

4 Convert to km: (a) 27 miles (b) 148 miles (c) 6.5 miles
5 Convert to miles: (a) 12 km (b) 345 km (c) 9.6 km

YARDS AND METRES

You may need to do these conversions when considering walking (or running) distances, or when measuring material.

The rough check
A metre is a tad bigger than a yard (actually, 1 metre = 1 yard and 3 inches). After converting yards to metres you should have a slightly smaller number.

The magic number
1 metre = **1.09** yards.

Benchmarks
100 metre = 109 yards
100 yards = 90 metres

Examples
Is a 200 metre race shorter or longer than a 220 yard race? By how many metres?

Rough check: coo, this is a tricky one to guess! A metre's longer than a yard, but is it long enough to make up the extra 20? Not sure. Probably not – it's only a tad longer.

Magic number: 200 × 1.09 = 218 and that's 2 less than 220, but you wanted the answer in metres, so let's do it the other way.

220 ÷ 1.09 = 201.83 so 220 yards is 1.83 m longer than 200 metres.

Practice 2

1 The last time they had new curtains in the Teach Yourself boardroom, it needed 28 yards of material. How many metres will they need for the new set?

(Contd)

2 Clare de Luney, the French champion racing snail can cover 1 metre in 9 minutes. Her trainer wants to enter her in the Frimley Gold Cup, a distance of 7 yards. The English champion, Slippery Sid can slither it in 56 minutes. Does Clare stand a chance?
3 Convert to metres: (a) 220 yards (b) 50 yards (c) 17.5 yards
4 Convert to yards: (a) 1000 m (b) 50 m (c) 12.4 m

FEET AND INCHES, CENTIMETRES AND MILLIMETRES

In the UK, we routinely use both Imperial and metric systems for measurements around the home and office, and around ourselves.

It makes sense to deal with these two pairs of units together. In practice you will very rarely convert a measurement solely in feet (sorry about the pun) into metric – it will normally be feet-and-inches, or just inches. The only difference between centimetres and millimetres, of course, is where you put the decimal point. It's a crucial difference, but easily handled.

Best foot forward...

If you are starting with feet and inches, the first job is to convert the measurement to inches (12 inches = 1 foot). For example:

3 feet 5 inches = 3 × 12 = 36 + 5 = 41 inches

If you started in metric and converted to inches, then at the end you need to divide this by 12 to break it into feet and inches.

67 inches ÷ 12 = 5 remainder 7 = 5 feet 7 inches

What about the little bits?

An inch is quite a large unit, and must be subdivided for any degree of accuracy. Tape measures are normally marked with 1/2, 1/4, 1/8 and 1/16 divisions. When converting inches to millimetres, you must first change these to decimals.

A4 paper is $8\frac{1}{4} \times 11\frac{11}{16}$ inches. What's that in mm?

$8\frac{1}{4} = 8.25 \times 25.4 = 210$ mm
$11\frac{11}{16} = 11.69 \times 25.4 = 297$ mm

The rough check
An inch is about 25 mm (millimetres). The easiest way to multiply by 25 is to multiply by 100 then divide it by 4 (or halve it twice).

6 inches = 6 × 100 = 600 ÷ 2 = 300 ÷ 2 = 150 mm

Similarly, when doing millimetres to inches, to divide by 25, multiply by 4 then divide by 100 (which you do by moving the decimal place back to stops).

750 mm = 750 × 4 = 3000 ÷ 100 = 30 inches

The magic number
1 inch = **25.4** mm. 10 mm = 1 cm. To change mm to cm, move the decimal point one place to the left. So, 1 inch = **2.54** cm.

Benchmarks
4 inches = 100 mm
1 foot (12 inches) = 300 mm

Examples
I have a section of skirting board 2.42m (2420 mm) long. Is it enough to fit across a wall 13 feet 2 inches wide?

Rough check: We could do 2400 ÷ 25, but that would be hard work. 2500 ÷ 25 is much easier, and that gives us 100. Then knock off a couple – call it 98 inches. Convert that into feet and inches: 8 × 12 = 96, which is near enough. The board is around 9 feet long, and it's way too short for the job.

Magic number: 2420 ÷ 25.4 = 95.3. 7 × 12 = 84, so 7 remainder 11.3.

2420 mm is 7 feet 11 inches

Practice 3

1. You have a 6 × 4 inch photo. Will it fit into a 100 × 160 mm frame?
2. Mac is 4 foot 5½ inches (sitting on a chair); Heidi is 6 foot 8 inches (standing on a chair). How much taller is Heidi, in centimetres?
3. Convert to mm: (a) 9 inches (b) 3 foot 3 inches (c) 2 ft 1¼ ins
4. Convert to (feet and) inches: (a) 300mm (b) 1570mm (c) 97 mm

---

Area

SQUARE MILES AND SQUARE KILOMETRES

These units are mainly used for expressing the size of geographical areas such as countries, lakes, national parks, big cities – all of which tend to be measured in the local system.

The rough check
1 square mile is just over 2.5 square kilometres. When converting square miles to square kilometres think 'twice as much and a fair bit more'; when going the other way 'quite a chunk less than half'.

The magic numbers
1 square mile = **2.59** square kilometres.

Benchmarks
10 square miles = 25 square kilometres
100 square kilometres = 40 square miles

Examples
The area of Wales is 8,021 square miles. Is it bigger or smaller than Belgium (30,051 square kilometres)?

Rough check: 8,000 × 2 and a bit is something over 18,000. Wales is probably a fair bit smaller.

Magic number: 8021 × 2.59 = 20,774. Wales is almost 10,000 square kilometres smaller than Belgium.

Practice 4

1 London covers 610 square miles. What's that in square kilometres?
2 The area of Canada is 3,855,100 square miles. The area of the USA is 9,826,630 square kilometres. Which is bigger?
3 Convert to square km: (a) 22 square miles (b) 1000 square miles
4 Convert to square miles: (a) 60 square km (b) 742 square km

SQUARE YARDS/FEET AND SQUARE METRES

The areas of floorings, flats, office space and the like may be expressed, in the UK, in either square yards, square feet or square metres. If you ever need to compare them, you need to have them all expressed in the same units.

The rough check

As we've already seen, a metre is a tad more than a yard, so a square metre is a tad more in both directions, or quite a bit more overall.

1 yard = 3 feet, so 1 square yard = 9 feet. And from those two we get:

1 square metre is about 10 square feet. That's going to give us nice easy rough checks.

The magic numbers

1 square metre = **1.195** square yards = you could normally work with **1.2**.

1 square metre = **10.76** square feet.

Benchmarks

100 square yard = 80 square metres
100 square feet = 9 square metres

Examples

You are looking for office space and find two possibilities, both equally suitable. The one in Bloomsbury will cost £55 per square foot, the one in Soho is priced at £625 per square metre. Which is cheaper?

Rough check: Working on 10 square feet = 1 square metre, the Bloomsbury office would cost around £550 a square metre, which makes it a good deal cheaper.

Magic number: 55 × 10.8 = 594. A bit more than the guesstimate, but still cheaper.

Practice 5

1. You have a lounge 13 foot 4 by 12 foot 9 (that's 170 square feet). How many boxes of laminate flooring will you need, if each can cover 2 square metres?
2. The rugby pitch at Toulouse has an area of 10,080 square metres. The football field at Chelsea is 8250 square yards. How much smaller is the Chelsea ground? Answers in square metres please.
3. Convert to square metres:
   (a) 144 square feet
   (b) 250 square yards
4. Convert to square feet:
   (a) 70 square metres
   (b) 438 square metres

Acres and hectares

Just in case you get into an agricultural situation and need these, here are the necessaries, in brief. Hectares are the bigger unit – over twice as big as an acre.

1 acre = 4840 square yards
1 hectare = 10,000 square metres
1 hectare = 2.47 acres
The magic number is 2.47

Weights and measures

POUNDS (AND STONES) AND KILOGRAMS

In the UK, most of us know our weight in stones and pounds; in the USA they have dropped the stones; most other places use kilos. If you start with stones and pounds, convert them to pounds first. (14 pounds = 1 stone)

The rough check
Kilos are bigger – at something over 2 pounds. When converting pounds to kilos, you finish up with a bit less than half.

The magic numbers
1 kilogram = 2.2 pounds.

Benchmarks
1 lb (pound) = ½ kg
1 st = 6 kg (and a bit)

Examples
After a good lunch, a typical publisher weighs 15 stone 3 pounds. What's that in kilograms?

Rough check: Call it 15 stone at 6 kg to the stone, and you get 100 kg. Or convert the weight to pounds: 15 × 14 = 210 + 3 = 213 and a bit less than half of that is about 100.

Magic number: 213 ÷ 2.2 = 96.8.

Practice 6

1. My cat weighs 17 lb. What's that in kilos?
2. A Smart car weighs 730 kg (before the driver gets in). What's that in pounds? And in stones and pounds?
3. Convert to kilograms: (a) 12 lb (b) 7 stone (c) 4 st 6 lb
4. Convert to stones and pounds: (a) 20 kg (b) 99 kg (c) 7.5 kg

OUNCES AND GRAMS

If you are a keen cook and like to explore recipes from around the world, you need to be proficient in switching between imperial and metric units. And those are not the only people who need to be able to convert between the two.

The rough check

An ounce may be a small measure, but grams are tiny.
1 ounce = a bit more than 25 grams.

The magic numbers

1 ounce = 28.35 grams.

Benchmarks

4 ounces = 100 grams

Examples

I have a Madeira cake recipe which calls for 175 g butter and of caster sugar, and 250 g self-raising flour. I want to give this to Aunty Flo who has scales marked in ounces. What are the weights in ounces?

Rough check: 175 ÷ 25 = 7; 250 ÷ 25 = 10. But it's a bit more than 25 grams to the ounce, so a bit less than 7 and 10 ounces.

Magic number: 175 ÷ 28.35 = 6.17; 250 ÷ 28.35 = 8.82 (tell her 6 and 9 ounces)

Practice 7

1. Which is heavier, a half pound (8 ounce) block of butter, or a 250 gram block? Answers in grams please.
2. Convert 1234 grams to pounds and ounces (16 oz = 1 lb).
3. Convert to grams: (a) 12 oz (b) 3 lb (c) 1 lb 4 oz
4. Convert to pounds and ounces: (a) 375 gm (b) 950 gm (c) 3210 gm

LIQUID MEASURES: GALLONS, PINTS, FLUID OUNCES, LITRES AND CC

Just so there's no doubt – we are talking about the gallons and pints they use in the UK and other parts of the world where Imperial measures hold sway – not the US version.

This section is for car drivers, cooks, and those drinkers that like to keep tabs on quantity (at least, for as long as they can count). The conversions are almost always between gallons and litres, or between pints and litres, or between fluid ounces and cc (cubic centimetres). You hardly ever get measurements in gallons and pints. When did you last hear anyone say, 'I must have had 1 gallon and 3 pints last night'?

What's in your pint?

Over in the US of A, pints are different. In fact, they have two different pints in the US – the ordinary (liquid) pint, which us 0.83 Imperial pints and the dry pint which is 0.97 Imperial pints. Dry pints? But then, Americans measure flour by the cup.

The rough check

A gallon is close to 4.5 litres, but you could call it 5.
A litre is about 2 pints.
A fluid ounce is about 30 cc.

The magic numbers

1 gallon = **4.55** litres.
1 litre = **1.76** pints.
1 pint = 20 fluid ounces
1 litre = 1000 cc
1 fluid ounce = 28.4 cc

Benchmarks

40 litres = 9 gallons (about a tankful in a hatchback or small saloon)
2 pints = 1 litre
10 fluid ounces = 300 cc

Examples

That Madeira cake recipe also needs 150 cc of milk. What's that in fluid ounces for Aunty Flo?

Rough check: 150 ÷ 30 = 5.

Magic number: 150 ÷ 28.4 = 5.28 (tell her 5 fluid ounces or ¼ pint)

Practice 8

1 Heidi's car will do 55 miles to the gallon, and the tank will hold 25 litres of petrol. Mac's moped will do 160 miles to the gallon and its tank will hold 7 litres of petrol. Which will go furthest on a full tank?

2 Half way through the 7th pint, Mac always starts to sing. On a trip to the Oktober Beerfest, can he get through 3 1-litre steins of beer quietly?

3 Convert to litres: (a) 12 pints (b) 20 gallons (c) 3.5 gallons

4 Convert to gallons:(a) 75 litres (b) 231 litres

5 Convert to pints: (a) 6 litres (b) 3.7 litres

What about mpg?

Though we've been buying petrol by the litre for years, most of us in the UK still think about fuel use in terms of miles per gallon (where anything over 50 mpg is good, anything under 30 mpg is a gas-guzzler). What do they do over there? Well, they don't use kilometres per litre, partly I suspect because the numbers can sound quite silly. Did you know that a Humvee does 3 kpl? Instead, they describe fuel use in terms of how many litres it takes to move the vehicle 100 kilometres. On that measure a Humvee uses 33 l per 100 km. Heidi's car (55 mpg) uses 8.3 l per 100 km.

TEMPERATURE

The Celsius system (which used to be called Centigrade) has taken over as the standard way to express temperature throughout the world – except for four oddities: Belize, Myanmar, Liberia and the United States. So, just in case you are planning a holiday in any of these places, here's how to convert Celsius to Fahrenheit.

The standard stress-free conversion routine won't work here, because the scales are a different length, and start at a different point.

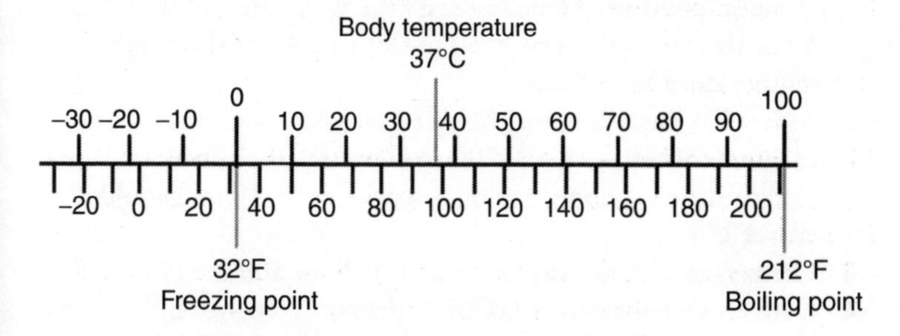

100° on the scale in Celsius equals 180° in Fahrenheit, or for every 10°C it goes up 18°F (or for every 5°C it goes up 9°F – remember those numbers). That's more or less double – but as the scales start at a different point, this won't work for rough checks.

The formula for converting °C to °F is: $F = C \times 9 \div 5 + 32$
Rough check: double it and add 30

And to convert °F to °C is: $C = (F - 32) \times 5 \div 9$
Rough check: subtract 30 and halve it

Benchmarks

0°C = 32°F	Freezing point (of pure water, at sea level)
100°C = 212°F	Boiling point (of pure water, at sea level)
37°C = 98.6°F	Normal body temperature
16°C = 61°F	Mild spring day

Examples

On holiday in Miami, the weather forecast said it would be around 85°. Will I need a warm cardie?

Rough check: 85 – 30 = 55. Half that is a bit more than 25.

Magic formula: 85 – 32 = 53 × 5 = 265 ÷ 9 = 29.4 (no cardie needed)

Practice 9

1. It's time for baby's first bath, and according to Spock (*Logical childcare*, Vulcan Press) the water should be 100.5°F. Your thermometer is marked in Celsius. What should the temperature be in °C?
2. Aunty Flo has asked for the secret of your delicious caramel toffees – which is that you must get the mix to exactly 124°C. Aunty Flo only works in °F. What should you tell her?
3. Convert to Celsius: (a) 98°F (b) 120°F (c) 300°F
4. Convert to Fahrenheit: (a) 10°C (b) 70°C (c) –40°C

CURRENCY CONVERSION

The standard stress-free conversion routine won't work here either, but that doesn't mean you have to get stressed about it. There are two tricks to avoiding stress. The first is to keep in mind the relative sizes of the units. If you are converting to or from UK pounds this part is very simple, because the pound is the biggest currency unit in the world (at the time of writing anyway). If you start with pounds you will always get more of the other currency, whatever it is – though not so many now (2010) than a few years back.

The magic numbers of exchange rates are changing constantly, so there is no point in putting them into a book – whatever we write now will be different tomorrow, let alone by the time this gets published and you are reading it. However, the principles are the same.

The same exchange rate can be expressed in four different ways. Let's look at the UK pound – euro rates. You might see:

£1.00 = €1.1089
€1.00 = £0.9018

Bank buys at: £1.00 = €1.15
Bank sells at: £1.00 = €1.05

You can use either of the first two for converting either way, multiplying or dividing as appropriate. That may sound woolly, but if you remember the thing about relative sizes, it's straightforward.

Converting UK pounds to euros, will give you a bigger number, so you can multiply by 1.1089 or divide by 0.9018 – both give the same answer.

£250   250 × 1.1089 = €277   250 ÷ 0.9018 = €277

Converting euros to pounds will produce a smaller number, so you need to divide by 1.1089 or multiply by 0.9018.

€500   500 × 0.9018 = £451   500 ÷ 1.1089 = £451

The second pair of figures, 'bank buys/sells at...' are there to remind you that banks and bureaux de change make money out of currency transactions by taking a little slice out of each deal.

Which brings us to the second trick of low-stress currency conversion: unless your name is George Soros, there's not a lot you can control here, so there's no point in getting uptight about it. Use one of the Web's many currency conversion tools (www.xe.com is easy to use, efficient and one of the shortest addresses you will ever have to type) and your sense of which is the largest unit, to get an approximate figure, and use this to check the amount that the bank or bureau gives you.

TIME ZONES

Before we leave this chapter, here are a few words about time, which is also different 'over there' (with a few exceptions).

If you are going abroad, or communicating with someone in another country, it's useful to know what time it is there in relation to the time where you are. Useful – and easily done.

Just check out a time zone map, and do the simple sum. The map opposite shows how far a selection of time zones differ from UTC (Coordinate Universal Time – the abbreviation comes from the French) – or GMT as it used to be called.

- Head East, towards India, China, Australia or wherever and they are further on in their day than we are, so you need to add the difference.
- Head West, towards the Americas and they are behind us, so you need to subtract the difference.

SOME BENCHMARKS TO REMEMBER

- Most of Europe – from Spain to Poland – is 1 hour ahead of UTC.
- India is 5½ hours ahead; China 8 hours ahead.
- The US East Coast is 5 hours behind; the West Coast is 8 hours behind – they are having breakfast in LA while you are wrapping up the job for the day.

AND A FASCINATING FACTOID:

- Russia is so big that Moscow is 3 hours ahead, but eastern Siberia is 12 hours ahead.

How many hours in the day?

This is not quite the silly question it may seem: the question is really, '24 or 2 lots of 12?' Over there they tend to use the 24 hour clock, and it does cut down the potential for confusion. You might, for example, set up an informal meeting with some French colleagues for 7 o'clock, intending a power breakfast and they would turn up for drinks and supper. To prevent this happening, you could specify 7 a.m. – and that would have exactly the same result because to a French speaker, 'a.m.' means 'apres midi' (after midday) not 'ante meridiem' (Latin, before noon). 7.00 hours is clearly different from 19.00 hours. Just remember to put your watch forward an hour when you cross the Channel.

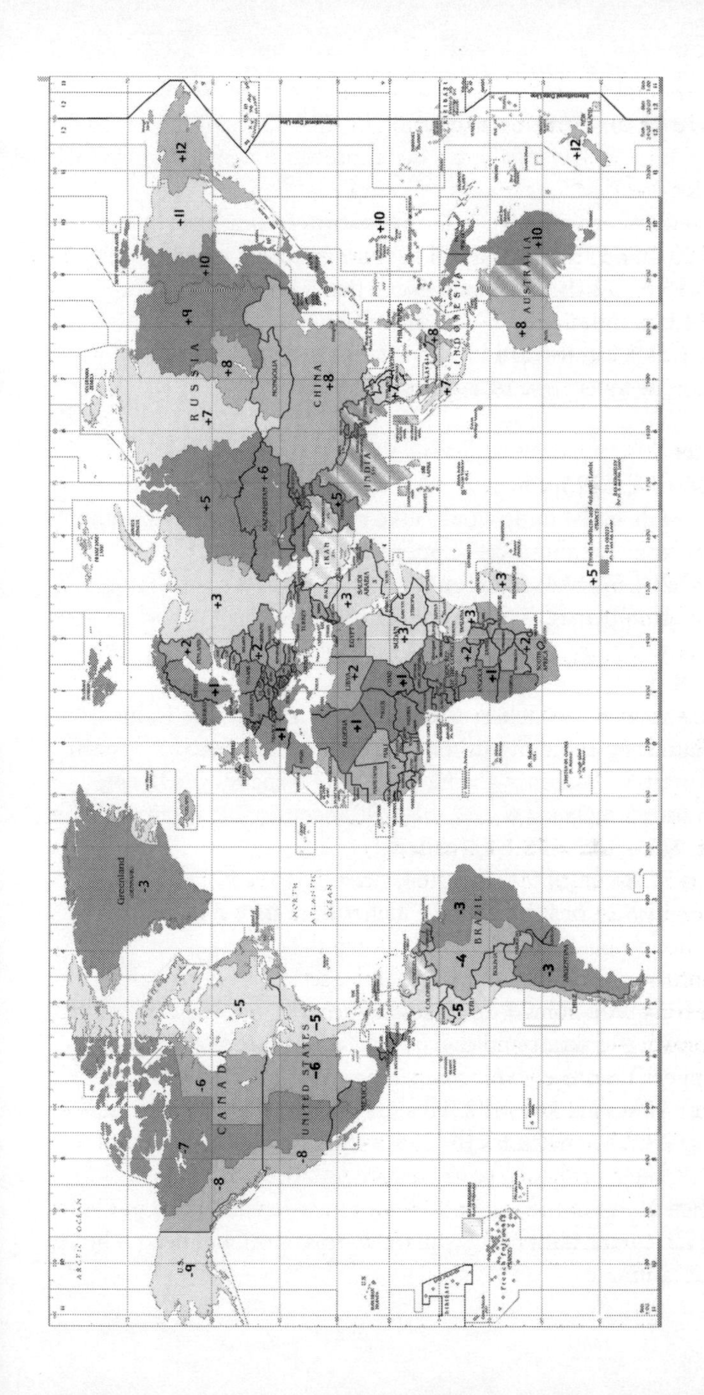

ARCTIC OCEAN
U.S.
-9
-8
-7
CANADA
-6
-5
UNITED STATES
-8
-6
-5
Greenland
-3
NORTH ATLANTIC OCEAN
BRAZIL
-3
-4
-5
-3
+1
+2
+3
+1
+2
+3
+2
+1
+2
+3
+5
+3
RUSSIA
+5
+6
+7
+8
+9
+10
+11
+12
CHINA
+8
INDIA
+5
+7
+8
+10
+7
INDONESIA
+8
AUSTRALIA
+10
+12

Answers and explanations

Practice 1

1 Heidi's car will have 120 miles worth of petrol at Le Havre. 120 m = 193 km. We will have to stop and refuel.
2 3.3 km = 2.05 miles, and that's just a bit too far.
3 24,901 miles = 40,074 kilometres
4 (a) 43.5 km (b) 238 km (c) 10.5 km
5 (a) 7.4 miles (b) 214.4 miles (c) 6 miles

Practice 2

1 28 yards = 25.6 m
2 7 yards is 6.4 m. It would take the French snail 57.6 minutes to cover the distance. Slippery Sid should win.
3 (a) 201 m (b) 45.7 m (c) 16 m
4 (a) 1094 yards (b) 54.7 yds (c) 13.6 yds

Practice 3

1 100 mm = 3.9 inches; 160 mm = 6.3 inches. The frame is more than long enough, but too narrow for the photo.
2 4 foot 5½ inches = 4 × 12 + 5.5 = 53.5 inches = 135 cm; 6 foot 8 inches = 6 × 12 + 8 = 90 inches = 228.6 cm. Heidi is 228.6 – 135 = 93.6 cm taller.
3 (a) 22.8 cm (b) 99 cm (c) 64.1 cm
4 (a) 11.8 inches (b) 5 foot 2 inches (c) 3 foot 2 inches

Practice 4

1 610 square miles = 1580 square km
2 3,855,100 square miles = 9,984,663 square km. Canada is bigger than the USA.
3 (a) 57 square km (b) 2590 square km
4 (a) 23.2 square miles (b) 286.5 square miles

Practice 5

1 170 square feet is 15.8 square metres. You will need 8 boxes of laminate.

2 8250 square yards is 6898 square metres. The Chelsea field is 10,080 – 6898 = 3182 square metres smaller than the Toulouse pitch.
3 (a) 13.4 square metres (b) 209 square metres
4 (a) 753 square feet (b) 4715 square feet

Practice 6

1 7.7 kg
2 730 kg = 1609 lb = 114 st 13 lb
3 (a) 5.4 kg (b) 44.4 kg (c) 28.1 kg
4 (a) 3 st 2 lb (b) 15 st 8 lb (c) 1 st 2.5 lb

Practice 7

1 8 ounces = 227 gm, so a 250 gm block is heavier, by 23 gm
2 1234 gm = 43.5 oz = 2 lbs 11.5 oz
3 (a) 340 gm (b) 1360 gm (c) 557 gm
4 (a) 13 oz (b) 2 lb 1.5 oz (c) 3210 gm = 113 oz = 7 lb 1 oz

Practice 8

1 25 litres = 5.5 gallons, at 55 mpg takes the car 302.5 miles.
2 7 litres = 1.5 gallons, at 160 mpg takes the moped 240 miles. Heidi's car goes furthest on a tankful.
3 3 litres = 5.3 pints, well below the singing point.
4 (a) 6.8 litres (b) 91 litres (c) 15.9 litres
5 (a) 16.5 gallons (b) 50.8 gallons
6 (a) 10.5 pints (b) 6.5 pints

Practice 9

1 100.5°F = 38°C
2 124°C = 255°F
3 (a) 36.7°C (b) 48.9°C (c) 148.9°F
4 (a) 50°F (b) 158°F (c) –40°F

5

Priceless percentages

In this chapter you will learn:

- ***Why and when these skills are useful***
- ***How to talk about percentages clearly***
- ***How to calculate percentage increases and decreases***
- ***How to estimate percentages***
- ***About VAT calculations***

Why and when these skills are useful

Percentages are used a very great deal in ordinary life because they are a very convenient and efficient way to express both fractions and multiples. They are used to describe interest rates on your savings, credit cards and mortgages, to set the amount of your annual pay award, to measure price increases and discounts in sales, to compare performance with targets, to describe changes over time and many, many other things. If you don't have a working grasp of percentages, you can get in a mess – and an expensive mess at that.

Bargain!

One of the inspirations for this book was a story in *The Guardian*. A man went into a gents' outfitters where they had a sale on – they were offering 25% off any shirt in the shop. He picked two, took them to the counter and the sales assistant said, 'There's 25% off, and you've got two, so that's 50% off.' As he left the shop, he wondered whether she would have given him 75% discount if he'd bought three, though perhaps if he had tried it with four shirts she might have seen the error in her arithmetic.

What's a percentage?

The clue is in the name: 'per cent' is Latin for 'in a hundred' or 'for every hundred'. So, '50 per cent' means '50 for every 100'. If a discount is 25%, it means that they will reduce the price by 25p in every £1 (100p). If your Council Tax has gone up 10%, then for every £100 you paid last year, this year you will be paying an extra £10.

In all these examples, you could have replaced the percentage with a fraction. 50% is a half; 25% is a quarter; 10% is one tenth. But percentages can also be used to express multiples. 200% means 200 for every 100, or twice as much; 1000% percent is 1000 for every 100 or 10 times as much. And this is where misunderstandings creep in – not because the numbers are misleading, but because people don't always express themselves clearly.

TALKING ABOUT PERCENTAGE INCREASES

These misunderstandings are most likely to occur when people are talking about percentage increases. For instance, what does this mean?

'The price of eggs has gone up 300% since last spring.'

Does this mean that eggs now cost three times as much – so if they were £1 last spring, they are now £3? Or does it mean that the increase in the price of eggs is 300%, i.e. £3 for every £1 they use to cost, so that they are now £4? Let's try it again with a smaller percentage, which most of us will find easier to visualize.

'The price of eggs has gone up 30% since last spring.'

That clearly means that there has been an increase of 30%, and that eggs are now £1.30.

If the phrase is 'gone up by X%' or 'increased by X%' or similar, then it should refer to the amount of the increase. Unfortunately, some people use that sort of phrase when they really mean that the new price is 'X% of what it was' previously. Unless you are their boss or their editor, there's nothing you can do to ensure that people express themselves clearly, but you can reduce the amount of confusion in the world by these two steps:

- Check the figures that the comparison is based on, and work out the percentage for yourself.
- Be clear in how you express yourself. If sales have increased from £100,000 to £250,000 then describe this either as 'an increase of 150%' or 'has increased to 250% of its previous value' – and in either case, give the actual figures. Alternatively – and this may be more effective in saving confusion – talk in terms of multiples, where the increases are more than 100%. 'Twice as much' is clearer than 'an increase of 100%' or '200% compared to its earlier value'.

IT DEPENDS WHICH WAY YOU LOOK AT IT

Here's another potential source of confusion with percentages – and again, the confusion does not come from the numbers, but from the way people talk about them. For example:

'There's a 20% difference between the profits we make on widgets and on doodahs...'

To make proper sense of this, you need to know what the 20% is based on.

Down at Krappat Manufacturing Co, the profit on widgets is £8000 a month. If you take '20% difference' to mean '20% more than on widgets', then the extra profit on doodahs is:

20% of £8000 = 20/100 × 8000 = 2/10 × 8000
= 2 × 8000 ÷ 10 = 16000 ÷ 10 = 1600

So the profit on doodahs is 1600 + 8000 = £9,600.

Except you then find that the actual profit was £10,000.
What went wrong?

Let's try it the other way, with '20% difference' meaning '20% less on widgets than on doodahs'.

20% of £10,000 (let's skip the intervening steps) = £2000

And doodahs make £2000 more profit than widgets.

People can be sloppy when they are talking about percentages – far more so than when using fractions. You won't hear someone say 'a fifth', by itself; it will always be 'a fifth of something'. But you will hear '20% out', 'a difference of 10%'.

If it's not clear what the percentage is based on, ask.

What's 15% of...?

The most common type of percentage calculation that most of us have to do is to find a given percentage of an amount: a 10% service charge on a bill, a 25% discount in a sale. These are straightforward to do, and there are some shortcuts that will make some calculations even simpler.

The rule is: multiply the base amount by the percentage number and divide by 100.

For instance:

5% of £30 = 30 × 5 ÷ 100 = 150 ÷ 100 = £1.50
25% of 80 = 80 × 25 ÷ 100 = 2000 ÷ 100 = 20

A percentage is, of course, a variety of fraction – though the denominator is always 100. Like other fractions, they can sometimes be simplified:

$$\frac{25}{100} = \frac{1}{4}$$

$$\frac{40}{100} = \frac{4}{10} = \frac{2}{5}$$

Which leads us to our shortcuts. If you want 25% of something, you can multiply it by 1/4 or – more simply – divide it by 4. Here are some shortcuts worth noting:

Percentage	*Fraction*	*Sum to do*
10	$\frac{1}{10}$	÷ 10
20	$\frac{2}{10} = \frac{1}{5}$	÷ 5
25	$\frac{1}{4}$	4
33⅓	$\frac{1}{3}$	÷ 3
40	$\frac{2}{5}$	× 2 ÷ 5
50	$\frac{1}{2}$	÷ 2
66⅔	$\frac{2}{3}$	× 2 ÷ 3 or ÷ 3 then subtract
75	$\frac{3}{4}$	× 3 ÷ 4 or ÷ 4 then subtract
90	$\frac{9}{10}$	÷ 10 then subtract

Try a couple:

Q The bill at the pizzeria was £32.50. How much is a 10% tip?
A £32.50 ÷ 10 = £3.25

Q What's 33⅓% of 450?
A 450 ÷ 3 = 150

It can sometimes be simpler to tackle the sum from the other end. Instead of calculating 75% of the value, calculate 25% and subtract that from the value – if you are working out a reduction, the subtraction is not needed. For example:

Q The office party Xmas lunch budget is to be cut by 75% next year. If it was £40 a head last year, how much will it be?
A If you cut something by 75%, you will have 25% left. 25% of £40 is £10.

The rough check

These shortcuts can also be used for rough checks, and you will need them when you are working out percentage sums on a calculator – as we are about to do.

Using a calculator

Once again, we have a very straightforward process, and made simpler on many calculators by the presence of a percentage key. Pressing the [%] key is equivalent to pressing [÷] [100] [=]. So, to find 30% of 250, you would do:

30 [×] 250 [%]

And that would give you 75. Let's run one more through the calculator, then it will be your turn. Rodger, of Dodger and Bodger the builders, is working out the cost of replacing the skirting board that your pet chipmunk ate when she escaped last week.

He's worked out it will be £375, for labour and materials. What will be the VAT at 17.5%?

Rough check: 17.5% is near enough to 20%, which is 1/5. If we reckon £400 ÷ 5, we get £80.

Calculated answer: 375 [×] 17.5 [%] gives 65.625 =.

£65.63 is in the same ballpark as £80, so that checks out.

Practice 1

1. I've ordered goods with a total cost of £568.50, but there will be a discount of 10%. What's the total cost?
2. In the USA, the standard tip rate is 15%. If the bill is $245.50, what should the waiter's tip be?
3. The stationery supplier offers a discount of 5% on bills over £100, or 10% over £500. I need £128 of paper, £275 of toner cartridges and £95 of envelopes. And I'm very tempted by a Parker ballpoint at £12. Should I add the pen to the order?
4. There's 25% off in the sale. What will two shirts, originally priced at £24.99 each, cost in total.
5. The departmental budget last year was £67,500. Staff costs took £59,800 of this. This year, the budget is to be increased by 6%, but staff costs will go up by 7.5%. Will there be more or less available for materials and other expenses?
6. Mac estimated that it would take him 5 days to write this chapter, but he was 20% out. How long did it take him?

What's this as a percentage?

In the second type of calculations, you need to express one amount as a percentage of another. For example, nationally the Fur-e-Nuff pet store chain spends 12% of its turnover on food. In the Didsbury branch, they have spent £17,500 on food against a turnover of £143,000. Is the manager over-feeding his stock?

To find what percentage 17,500 is of 143,000, we need to divide – which gives us the fraction, then multiply by 100 to get the percentage. Like this:

Rough check: 17,500 rounds to 20,000; 143,000 rounds to 140,000

$20{,}000 \div 140{,}000 = 2 \div 14$ (after knocking off zeros from both sides)

$2/14 = 1/7 \times 100 = 100/7$

$100 \div 7 =$ too difficult, round 7 to 10
$100 \div 10 = 10\%$

Calculator answer: 17,500 ÷ 143,000 [%] gives 12.27%

Notice the rough working. You have to reduce the values to nice round numbers if you want to work with them comfortably.

A second example would be where you wanted to know the percentage increase in sales, comparing one year to the next. Last year, your branch sold £547,000 of stuff, this year the sales were £582,500. What's the percentage increase? Here's one way to tackle it:

1. Find the actual increase: £582,500 – £547,000
   Rough check: 580,000 – 550,000 = 30,000
   Calculator answer: 35,500

2. Divide the increase by the base amount:
   Rough check: 30,000 ÷ 550,000 = too difficult,
   30,000 ÷ 600,000 = 3 ÷ 60 (knock off the spare zeros)
   = 1/20 = 5% on the shortcut table
   Calculator answer: 35,500 ÷ 547,000 [%] gives 6.49%

If the size of the increase is more than the original amount, then the result will be over 100%, but that's OK – we can use percentages to express multiples. For instance, a house bought in 1995 for £105,000 was sold in 2010 for £245,000. What was the percentage increase in price? We use exactly the same technique:

1 Find the actual increase: £245,000 – £105,000
Rough check: 250,000 – 100,000 = 150,000
Calculator answer: 140,000

2 Divide the increase by the base amount:
Rough check: 150,000 ÷ 100,000 = 150 ÷ 100 (knock off the thousands)
= 150% on the shortcut table
Calculator answer: 140,000 ÷ 105,000 [%] gives 133.33%

Decreases

You can work out a percentage decrease in almost exactly the same way. Last year the firm spent £15,689 on electricity. With the new energy-saving policy, this year's total bills came to £14,327. By what percentage has the bill gone down?

1 Find the decrease: 15,689 – 14,327 = 1,362
Rough check: 16,000 – 14,000 = 2,000 (but it was a good chunk less than 16,000 and quite a bit more than 14,000 so the result should be a fair whack less than 2,000)

2 Divide the decrease by the base amount:
Rough check: 2,000 ÷ 16,000 = 2 ÷ 16 (knock of the spare zeros)
= 1/8 × 100 = 100/8 = 12 and a bit (12 times table)
12%
Calculator answer: 1,362 ÷ 15,689 [%] gives 8.68%

Practice 2

1 You've been asked for a list of employees broken down by age and sex (a description which clearly applies to Geraldine, Frank and Phil). In terms of numbers, there are 49 people in total. Of these 32 are female, and 17 are over 50. What percentage are male? What percentage are over 50?

2. Heidi's car weighs 470 kg empty, but 850 kg when she, Mac, Rusty the dog and their luggage are on board. The car can carry up to 80% of its weight. Is it overloaded?
3. In our industry, turnover has fallen in the recession by an average of 8%. Our turnover dropped from £875,050 to £794,000. Are we doing better or worse than average?
4. The average salary in Phatkhat Inc is £25,468. The CEO's salary is £426,932. What is that as a percentage of the average?
5. On the Krappat Manufacturing assembly line, they aim for a maximum of 1.5% faulty units. If 7,450 widgets are produced in a day, and 134 are faulty, is this within the acceptable limits?

That's interesting...

Interest rates are always calculated in percentages. If you are borrowing or saving money, failure to understand this aspect of percentages can be an expensive shortcoming. We will be looking at interest rates when we turn to personal finance in Chapter 10.

VAT

Those goods which are subject to Value Added Tax may have their prices quoted VAT inc. or exc. VAT – including or excluding VAT. Why? Because if you are buying for a VAT-registered business, the VAT is largely irrelevant, as you will reclaim it from the VAT man. With stuff that is only likely to be bought by people for their own use, like clothes and DVDs, the prices rarely even bother to mention VAT (except in very small print for legal reasons) because it is unavoidable. With stuff that is only likely to be bought by businesses, like raw steel, private jets and anything in bulk, the prices are normally given without VAT. But it varies with the in-between stuff, such as stationery, car hire and all sorts of services.

E-shopping adds a second angle to the inc. or exc. VAT question. UK-based companies charge VAT on their sales, but if the goods are to be exported then either the VAT isn't charged or it can be reclaimed by the buyer. And companies based outside the EU tax area, but selling into the UK, don't normally charge VAT, though import tax may be due.

All of which means that it's worth checking the real cost of goods to you.

How does VAT work?

The theory of VAT is that the tax is due on the value added to goods as they are manufactured, processed, delivered, etc. The way the tax is collected takes a moment to explain so pull up a chair – and we'll find out how much tax was paid on it and by whom. (For the sake of simplicity, the VAT rate is 10% in this example.)

Step 1: The lumber company sells £100 of wood to the chair-maker, and charges £10 VAT on it. The company sends the £10 (along with the rest of the VAT it has collected) to the VAT man.

Step 2: The chair-makers turns the wood into chairs, which they sell to the shop for £500, plus £50 VAT. They keep back the £10 VAT they paid and send the remaining £40 to the VAT man.

Step 3: The shop sells the chairs to the public for £1000, plus £100 VAT. They keep the £50 VAT they paid and send the remaining £50 to the VAT man.

At the end of the process, the VAT man has £100 and that has come from the end buyer.

In the USA, they don't have VAT. Instead they have retail tax, which is paid by people when they buy stuff from shops. Which means that the tax man still gets his cut, but only the shop keepers have the bother of collecting it.

CALCULATING VAT – BOTH WAYS

If you are given the price exclusive of VAT, then calculating the tax is easy – it's just like working out a percentage increase. At the time of writing, the standard rate of VAT is 17.5%, so the VAT on goods worth £50 (exc.) is:

£50 × 17.5%

Rough check
17.5% is nearly 20%, so divide by 5
£50 ÷ 5 = £10 but it will be a bit less than this.

Calculator
50 [×] 17.5 [%] gives 8.75

It's a tad trickier the other way round. If Heidi bought a pen for £9.99 inc. VAT, how much of that was tax?

£9.99 is a nasty number, let's try another one.

If it cost £117.50 inc. VAT, how much was the tax?

I hope you spotted that this breaks into £100 cost + £17.50 VAT. The simplest way to deal with this sort of calculation is to realize that the VAT inclusive price is 117.5% of the base price (... while VAT is 17.5%). And we can get the base price by dividing the VAT inc. price by 117.5%: not the sort of sum to do in your head, but simple enough on a calculator. The VAT is then the difference between the two. Back to Heidi's pen:

£9.99 ÷ 117.5%

Rough check
If you divide by a bit more than 100%, that's like dividing by a bit more than one, so the answer will be a bit less than what you started with. And that's near enough for now.

Calculator
9.99 [÷] 11.75 [%] gives 8.50

So the VAT was 9.99 – 8.50 = 1.49

And, of course, you could also work out the VAT, as before, by multiplying the base cost by 17.5%.

Practice 3

In these questions, assume the VAT rate is 17.5% unless it says otherwise.

1. The on-line car hire company quotes you £50 a day, but that doesn't include VAT. What is the actual cost to you?
2. The Crap At Numbers Telephone Helpline Inc. is VAT registered. The phone bill last month was £176.35, including VAT, and the electricity bill was £94.75 including VAT. (VAT is charged at 5% on gas and electricity.) What was the total VAT on the two bills?
3. Here at CANTH Inc we charged £4,287.50 in VAT last month. How much was our total income?

Answers and explanations

Practice 1

1. 10% of £568.50 = 1/10 × 568.00 = move the decimal point one place left.
   = £56.85 and that's the discount
   568.50 – 56.85 = £511.65

2. 15% of $245.50 =
   Rough check
   10% of $245.50 is about $25, and 15% is half as much again. Half of 25 is about 12, so the tip should be around $37.

   Calculator
   245.5 [×] 15 [%] gives 36.825, which rounds up to $37

3 £128 of paper, £275 of toner cartridges and £95 of envelopes make a total bill of:
Rough check
100 + 300 + 100 = 500 – it's going to be close to the £500 higher discount level

Calculator
128 + 275 + 95 = 498
At £498 the discount is 5% = £24.90, so the bill would be £473.10
Add in the £12 pen, and the total is £510. The discount is now 10% = £51.00. This makes the total bill £459.00. Buying the pen reduces the cost by about £14.

4 Get the total cost first, then take off the 25% discount.
2 × 24.99 = 49.98 (total original cost)
25% of £49.98 = 1/4 × 49.98 = 49.98 ÷ 4 = 12.495 = £12.50
£49.98 – £12.50 = £37.48

5 Last year: budget = £67,500, staff costs = £59,800, other costs = 67,500 – 59,800 = 7,700
This year, budget = 67,500 + 6% of 67,500 = 67,500 + 4,050 = 71,550

staff costs = 59,800 + 7.5% of 59,800 = 59,800 + 4,485 = 64,285
remainder = 71,020 – 64,285 = 7,265.
There is less available for other costs this year.

6 20% out is not very helpful. It could mean 20% more or 20% less. The answer is either 6 days or a bit over 4 days.

Practice 2

1 Age: 17 out of 49 as a percentage
Rough check
20 ÷ 50 × 100 = 2 ÷ 5 × 100 = 200 ÷ 5 = 40

Calculator
17 [÷] 49 [%] gives 34.7%

Sex: 32 out of 49 as a percentage

Rough check
30 ÷ 50 × 100 = 3 ÷ 5 × 100 = 300 ÷ 5 = 60

Calculator
32 [÷] 49 [%] gives 65.3%
That's the females, so the male percentage is
100 – 65.3 = 34.7%

2 Weight being carried = 850 – 470 kg
Rough check
That's about 400 kg. The car weighs about 500 kg and 80% of this is 4/5 × 500 = 400. It must be close to the limit.

Calculator
850 – 470 = 380
470 [×] 80 [%] gives 376
The car is carrying 4 kg more than its safe limit. Rusty will have to run alongside.

3 A fall of £875,050 to £794,000 is £81,050. This is a bit less than 10% of £875,050. We are probably doing worse than average. On the calculator:
81,050 [÷] 875,050 [%] gives 9.26%

4 426,932 as a percentage of 25,468 is going to be in the thousands. 500K compared to 25K = 500 ÷ 25 = 20 times as much = 2,000%
Calculator
426,932 [÷] 25,468 [%] gives 1,676%
But people will understand the difference better if this is expressed as 16.76 times as much.

5 134 out of 7,450 =
Rough check
1% of 7,000 = 70; 0.5% = half of that = 35, so 1.5% = something over 105. 134 is probably too high.

Calculator
134 [÷] 7,450 [%] gives 1.79%: time to overhaul the production line.

Practice 3

In these questions, assume the VAT rate is 17.5% unless it says otherwise.

1 £50 × 17.5% = £8.75. The total cost is £58.75

2 Phone bill ex. VAT = £176.35 ÷ 117.5% = £150.08;
VAT = £26.26
Electricity bill ex. VAT = £94.75 ÷ 105% = £90.23;
VAT = £4.51
Total VAT = £26.26 + £4.51 = £30.77

3 To calculate VAT, you multiply the income by the VAT rate. To get the income from the VAT we need to reverse the calculation.
Rough check
4,000 ÷ 20% = 4,000 ÷ 1/5 = 4,000 × 5 = 20,000

Calculator
£4287.50 [÷] 17.5% [%] gives £245.00
(A nice little earner – shame it's fictional.)

6

Numbers at work: reading a spreadsheet

In this chapter you will learn:

- ***Why and when these skills are useful***
- ***Some spreadsheet basics, and why they matter***
- ***How to make sense of budgets***
- ***About 'what-if' forecasting and the unknown future***
- ***How to set up running totals and running averages***
- ***How to identify trends and to project them***

Why and when these skills are useful

Note the title – this chapter is mainly about *reading* spreadsheets, not writing them, though we will do a little of that. We are assuming that you are likely to be faced with spreadsheets that others have set up, but which you have to make some sense of, and into which you may have to enter some data or calculations of your own.

To use a spreadsheet efficiently, you need to understand what it will do and what it won't do – and it helps if you have a grasp of how it does things. When you are entering data into a sheet, you need to know what you can change and what you should leave well alone. When you are trying to draw meaning from a sheet, you need to know which values you can rely on and which should be taken with a pinch of salt. And that is what this chapter is about.

New to spreadsheets?

If you know nothing about spreadsheets, there's a good short introduction to them in *Get Started in Computing*, or for a more thorough introduction you might like to try *Get Started with Excel.*

Spreadsheet basics, and why they matter

We're assuming that you have done some work with spreadsheets in the past, and that you know about rows and columns, how to enter data and stuff like that. However, we are going to lay out a few of the basic principles here because of the impact they have on how you use and view a spreadsheet. So, while it may seem like we are stating the bleedin' obvious, bear with us because it matters.

A spreadsheet cell can contain:

- **Numbers**, which are held accurate to umpteen decimal places, but which can be displayed as currency, percentages, whole numbers and various other ways. So, what you see is not necessarily what's actually in the cell, but may be rounded up or down to some extent.
- **Formulae**, which perform calculations using values which are either written directly into the formula or drawn from the contents of other cells. These formulae can handle very complex calculations using ready-made functions.
- **Text**, so that you can write headings onto columns and rows, or put notes and labels beside the formulae so that anyone reading the sheet can understand what it is about.

A blank spreadsheet has no structure. Any and every cell can be used in exactly the same way. The upshot of this is that the layout of a sheet depends entirely on the person who builds it. It will typically reflect the way that data would have been collated on paper. There are certain conventions. You will normally find these features:

- Sheets are given a (descriptive) title, and this is written in the top left cell (A1), or is centred across the top row of the first few columns.
- Totals are put at the bottom of columns or to the right of cells, but they could be at the top or left, or anywhere else on the sheet.
- Related sets of data, such as the results by category for different years or different departments are written in adjacent columns or rows. This makes it easier to compare one item with another.
- Blank lines are left between blocks of data, if they are not related.

Accurate does not mean correct

A spreadsheet can calculate an answer to 15 significant figures, which could describe the distance from here to the moon accurate to 1/1000 of a millimetre. But that does not mean that the answer is correct. Any results produced by a spreadsheet are only as good as the data that went into it, and the formulae that are working on it. You need to keep this in mind, always.

Making sense of budgets

If you go into 'non-numerate' careers, like the law, social work, teaching (except for science and maths), or many aspects of the civil service, you may be able avoid any real contact with numbers until you start to get into management and are presented with your first budget. Suddenly you have to work out what it means and how much money your section has got for what sort of things – or worse, you have to work out how to cut costs to match a smaller allocation. Don't panic. These things do make sense, once you know how to look at them. The examples that follow are simplified versions of what you might see in a public sector organization, or a department within a larger company – in both cases the income is set by a higher level of management, and not generated by the organization (except perhaps for some limited sales of services, sub-letting of facilities, etc.).

VARIANCE

Let's start with a budget and its outturn, and compare actual spending and income over the year with the budget plan. Layouts vary hugely, but you should find a column marked 'Variance' or something similar, which shows the difference between budget and actual values. There should also be a percentage variance column (see page 94).

Percentage variance can be more meaningful than the number – if there's an overspend of £1,000 on both salaries and stationery, but the salary budget was £500,000 and the stationery budget was £500, then you need to find out what's happening in the stationery cupboard. As a general rule of thumb, if the actual is no more than around 3% out (above or below) the budget figure in any category, then your budgeting and financial control are pretty good. In the example given here, either the managers were unrealistic about costs when setting the budget, or they have not kept costs under control during the year, or external events have created unforeseen costs.

As a manager you need to understand what caused any variance from your budget, in any category, but you should only spend significant time and effort on finding the causes of variance over the 3% threshold. What could have been foreseen by a better analysis of the situation at the time of setting the budget? What changes were unforeseeable? Can more flexibility be built into the next budget to handle the unforeseeable?

Budget 2009–10	2009–10 budget	2009–10 actual	Variance	Variance %
Salaries				
Front line staff	500,000	546,000	46,000	9%
Back office	110,000	123,000	13,000	12%
Ancillary	67,500	65,000	–2,500	–4%
Temporary	20,000	28,500	8,500	43%
Total salaries	697,500	762,500	65,000	9%
Premises				
Rent	15,000	15,000	0	0%
Heat and light	4,000	4,500	500	13%
Maintenance	1,000	1,250	250	25%
Cleaning	3500	4,000	500	14%
Total premises	23,500	24,750	1,250	5%
Other expenses				
Stationery	2,000	2,700	700	35%
Small equipment	2,500	3,300	800	32%
Advertising	1,000	975	–25	–3%
Total expenses	5,500	6,975	1,475	27%
Total outgoings	726,500	794,225	67,725	9%
Allocation	720,000	720,000	–	0%
Self-generated	30,000	25,700	4,300	14%
Balance b/f	5,400	5,400		
Total available	755,400	751,100		
Balance	28,900	–43,125		

BUDGET SETTING

The first stage of setting a budget is examine the previous year's outturn, and see what lessons are to be learnt from that. The next stage is to see what income you have in the coming year and match that to expenditure. You might start with something like the example shown here.

This gives the previous year's budget (Column B), then initial values for this year. The initial outgoing values were here produced by adjusting for inflation (Column C), though the income values were arrived at a different way. The allocation was fixed by those higher up the power ladder, the self-generated income figure was carried over, and the balance b/f (brought forward) is the overspend from last year.

As the manager, you will be expected to match your outgoings with your income. In good years, when there is more than enough income to meet existing commitments, this is an opportunity to invest and to improve provision. When money is tight, you have the far harder task of finding ways to cut costs and/or increase income. In the example shown here, the initial budget figures show an overspend of over £40,000. That must be reduced to zero. So how should you go about setting your budget in a time of cuts?

The first thing is to take straight across from C to D those things that cannot be changed, or cannot be changed easily, e.g. rent and salaries of permanent staff. If people are leaving and it is possible to manage without them, then their salaries can come out at this stage.

Next, look for possible savings, such as restricting the use of temporary staff, energy saving measures, etc. Always tackle the easiest ones first. It's good psychology because you can feel that you are getting somewhere, and it reduces the amount that you have to cut elsewhere. But be realistic. It's no good planning to cut temporary staff costs by 25% if in practice you will have to employ more than that to cover essential services.

MANAGING THE BUDGET

Having set the budget, you need to keep an eye on it over the year. One simple way to do this is to look at the figures every month,

A	B	C	D	E
Budget 2010–11				
	2009–10	Inflation adjusted	2010–11	
Salaries		2.50%		
Front line staff	546,000	559,650	559,650	
Back office	123,000	126,075	116,000	#1
Ancillary	65,000	66,625	66,625	
Temporary	28,500	29,213	25,000	#2
Total salaries	762,500	781,563	767,275	
Premises		5%		
Rent	15,000	15,750	15,750	
Heat and light	4,500	4,725	4,000	#3
Maintenance	1,250	1,313		
Cleaning	4,000	4,200		
Total premises	24,750	25,988	19,750	
Other expenses		5%		
Stationery	2,700	2,835		
Small equipment	3,300	3,465		
Advertising	975	1,024		
Total expenses	6,975	7,324		
Total outgoings	794,225	814,874	787,025	
Allocation	750,000	760,000	760,000	
Self-generated	25,700	25,700	35,000	
Balance b/f	5,400	–13,125	–13,125	
Total available	781,100	772,575	781,875	
Balance	–13,125	–42,299	–5,150	

#1 P/t clerk leaving: do not replace
#2 Contain the use of temps
#3 Energy-saving practices

comparing the amount that should have been spent up to that point with the actual expenditure. So, if the accounting year starts in April then at the end of June you should have spent 3/12 (25%) of the budgeted amount under each heading.

REVIEWING THE BUDGET

A	B	C	D	E	F
Budget 2009–10, monthly review October, Month 7 Ratio 7/12 = 58%					
	2009–10 budget	Budget to date	Actual to date	Variance	Variance %
Salaries					
Front line staff	500,000	291,667	318,500	26,833	9%
Back office	110,000	64,167	71,750	7,583	12%
Ancillary	67,500	39,375	37,917	–1,458	–4%
Temporary	20,000	11,667	16,625	4,958	43%
Total salaries	697,500	406,875	444,792	37,917	9%
Premises					
Rent	15,000	8,750	8,750		0%
Heat and light	4,000	2,333	2,625	292	13%
Maintenance	1,000	583	729	146	25%
Cleaning	3,500	2,042	2,333	292	14%
Total premises	23,500	13,708	14,438	729	5%
Other expenses					
Stationery	2,000	1,167	1,575	408	35%
Small equipment	2,500	1,458	1,925	467	32%
Advertising	1,000	583	569	–15	–3%
Total expenses	5,500	3,208	4,069	860	27%
Total outgoings	726,500	423,792	463,298	39,506	9%
Allocation	720,000	420,000			
Self-generated	30,000	17,500			
Balance b/f	5,400	3,150			
Total available	755,400	440,650			
Balance	28,900		–22,648		

The example here shows the state of play at the end of October. That's 7/12th of the year, so the ratio is 58.3%. Column C has the budget figures times 58.3%; Column D has the actual spend to date; in E and F we see the variance numbers and percentages. The manager should be doing the same thing here as in the budget outturn at the start of this section – where the variance is more than a few per cent out, you need to know why.

Good budgetary control will spot problems as they arise, solve them if possible, or arrange additional finances if needed to cope with unavoidable overspends.

Play with the sheets

The Excel spreadsheets used for these examples are available online (see page vii).

What-if forecasting

In the dim and distant past, in the years BC (before computers), if business people wanted to assess the possible profitability of a new product, or predict their cash flows over the coming year, they would get out a sheet of paper, a pencil and a hand-cranked adding machine (or an abacus), write down their best guesses as to what might happen, then work out the effects on the finances of the company. It all took time and effort – so much so that would normally only bother to work out best and worst case scenarios, or calculate the impact of only the most likely possibilities. They would work with rounded figures, because they knew some of the critical ones were only guesses, and they would look at the rounded numbers which came out of the process and recognize them for what they were – ballpark figures, based on guesswork, for guidance only.

Nowadays, you can set up a spreadsheet to analyse profitability and cash flows (and much else besides), type in your known values and estimates and get the answer instantly. You can then, in a matter of seconds, or at most minutes, rework the calculations with different estimates. In as long as it once took to produce a single round-number forecast, you can now have two dozen accurate to 3 decimal places forecasts – and every one of them will be exactly wrong. Let's see where they go wrong.

CASH FLOWS

Unless a business is sitting on a nice wedge of cash and has a steady income stream and no unusual costs in the offing, it needs to forecast its cash flows for the months ahead so that it can arrange bank loans, if necessary, or take steps to avert trouble.

Take a look at these two tables (pages 100 and 101). The first holds 'accurate' figures. Known values are typed in as exactly as possible – rent, premises costs, salaries, and similar can be known fairly accurately in advance. The cost of purchases can be calculated from the sales – for simplicity they have been set here at 50% of the sales for the month – as can commission. The sales are the big unknowns here, as in most companies, and the figures have been generated by a formula that uses the random number function. This is a valid way to do it. The sales and marketing people should be able to come up with a range of possible sales figures, e.g. between 20,000 and 25,000, or 500 +/– 50. If you were doing this by hand, you would probably take the median value; with a spreadsheet you can produce a random number within the range.

Random numbers in Excel

The RAND() function produces a random number between 0 and 1. By adding in a bottom limit and a multiple factor this can produce a random number in any range. So, to get a number between 20,000 and 25,000 you would use this formula:

```
= 20000 + RAND() * 5000
```

RAND() is recalculated every time the spreadsheet is updated, i.e. whenever new data is added or anything is changed, or when the F9 key is pressed.

THE ‘ACCURATE’ FIGURES

A	B	C	D	E	F	G	H
		Jan	Feb	March	April	May	June
Sales							
Shop		15943.29	23934.28	11847.77	11679.34	24382.55	24933.73
Online		13141.26	18785.40	10214.31	23977.49	21368.66	16414.15
Total		29084.55	42719.67	22062.07	35656.83	45751.21	41347.87
Purchases		14542.27	21359.84	11031.04	17828.41	22875.60	20673.94
Expenses		4126.00	4126.00	4126.00	4126.00	4126.00	4126.00
Salaries		11879.00	11879.00	11879.00	11879.00	11879.00	11879.00
Commission		318.87	478.69	236.96	233.59	487.65	498.67
Premises		1978.00	1978.00	1978.00	1978.00	1978.00	1978.00
Total outgoings		32844.14	39821.52	29250.99	36045.00	41346.26	389155.61
Monthly profit/loss		–3759.59	2898.15	–7188.92	–388.17	4404.95	2192.26
Bank balance	–5000	–8759.59	–5861.44	–13050.36	–13438.53	–9033.58	–6841.32

THE ROUNDED FIGURES

A	B	C	D	E	F	G	H
		Jan	Feb	March	April	May	June
Sales							
Shop		15900	23900	11800	11600	24300	24900
Online		13100	18700	10200	23900	21300	16400
Total		29000	42600	22000	35500	45600	41300
Purchases		14500	21300	11000	17800	22800	20600
Expenses		4100	4100	4100	4100	4100	4100
Salaries		11800	11800	11800	11800	11800	11800
Commission		300	400	200	200	400	400
Premises		1900	1900	1900	1900	1900	1900
Total outgoings		32600	39500	29000	35800	41000	38800
Monthly profit/loss		–3600	2900	–7000	–300	4600	2200
Bank balance	–5000	–8600	–5500	–12500	–12800	–8200	–6000

That first table shows that in April the company is going to need an overdraft of £13,438.53, but that this will drop to £6,841.32 by the end of June. Someone reading this sheet might be inclined to round those up to £14,000 and £7,000 – but this is still more accurate than can be justified by the sheet. Those pennies at the end of the figures have given them a spurious sense of accuracy.

In the second table, the values have all been rounded down to the nearest 100. It is now obvious from the look of them that these are not accurate, and they are going to be treated with much more circumspection.

Knowing the future

If you want to know exactly what will happen in the future, you will have to wait. Anything else is guesswork, no matter how complex the formulae that are used to predict it.

BREAKEVEN POINT

When a company is developing a new product, the key question is 'Can we make a profit from it?' And following on from that, 'How many do we have to sell, at what price, before we break even?' The breakeven point is reached when the profit from sales covers the initial development costs – design, tooling-up, training, launch, etc.

There are two main variables here – the sale price and the number sold. (In practice, there are variables in the initial costs as well, and there may be room for manoeuvre on the unit cost, but we will ignore these.) In the example sheet on page 103, these have been handled by setting a range of possible sales, from 2,000 to 20,000, with the price written into a cell. The sales income values are produced by multiplying the sales figures by the sale price, using the cell reference. A whole new set of results can be produced instantly by writing a different price into that cell.

Graph it!

We are trying to see the point at which the profit on sales exceeds the initial costs. Running your eye along the profit line, you can see that

BREAKEVEN POINT

A	B	C	D	E	F	G	H	I	J	K
	Unit cost	£3.50	Sale price	£7.99						
Sales	2000	4000	6000	8000	10000	12000	14000	16000	18000	20000
Cost	7000	14000	21000	28000	35000	42000	49000	56000	63000	70000
Income	15980	31960	47940	63920	79900	95880	111860	127840	143820	159800
Profit	8980	17960	26940	35920	44900	53880	62860	71840	80820	89800
Initial costs	50000	50000	50000	50000	50000	50000	50000	50000	50000	50000

this goes over £50,000 between columns F and G, and looking up from there we see that the sales are around 12,000 at that point. A graph can often make things clearer. And now you might see why we have written the initial costs in as a line of data. By graphing them and the profit figures we get a nice clear image of the breakeven point on our graph.

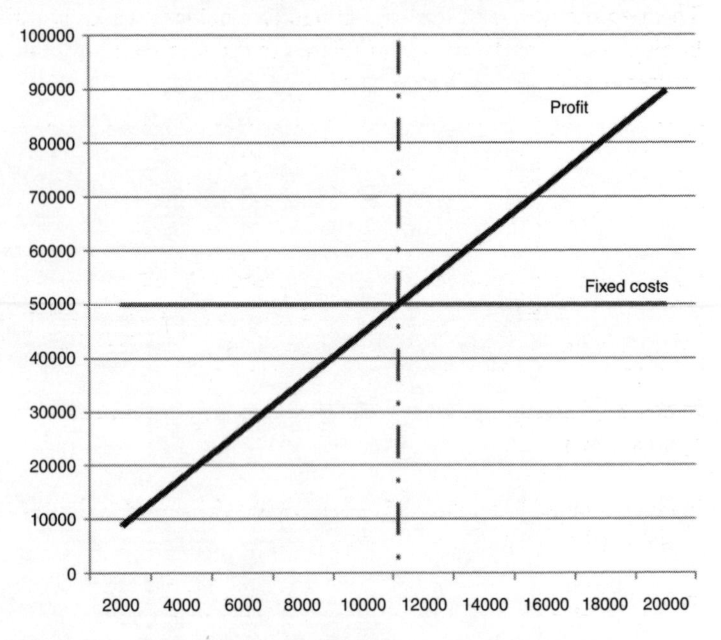

BODMAS on spreadsheets

Spreadsheets follow the BODMAS order, and you need to bear this in mind when writing formulae. See Chapter 1, page 17 for more on BODMAS.

Running totals and running averages

RUNNING TOTALS

Running totals can be very useful for following progress over time, or across a range of sales, component costs or similar. They give you a clear view of how close you are to your targets, or to the budget limitations. Running totals are easily set up – all you need is a simple formula to add an item into the total so far, then copy it down (or across) the set.

Through the power of our crystal ball (see below), we know this book's sales over the next year. Here are the first sixth months, written into a spreadsheet:

	A	B	C	D	E	F	G
1	Crap At Numbers sales						
2							
3		July	Aug	Sept	Oct	Nov	Dec
4	Month sales	3000	1250	2500	6500	1000	10250
5	Running total	3000	4250	6750	13250	14250	24500

In cell B5 is the simple reference '=B4' to copy the first month's figures down.

In cell C5 is the formula '=B5 + C4' to add the month's figures to the total.

This is copied across so D5 holds '=C5 + D4', and so on.

So that's 24,500 sales by Christmas. This year we can afford a turkey *and* the stuffing!

RUNNING AVERAGES

If you are looking for patterns in sales or costs or production levels or failure rates, or any other series of values, the raw data can make it difficult to see. Numbers can vary in response to random factors – the weather, holidays, machine failure, shortages of material, World Cup matches during working hours, the health of key workers, etc.

If you look at each month's sales of the book, you will see that they are all over the place, making it hard to tell how it is doing in general.

But if we calculate the average sales for the previous three months, and carry that across, this will smooth out the peaks and troughs and gives us a clearer view:

	A	B	C	D	E	F	G
1		July	Aug	Sept	Oct	Nov	Dec
2	Month sales	3000	1250	2500	6500	1000	10250
3	3 month average			2250	3417	3333	5917

In cell D3 is the formula '=(B2+C2+D2)/3' which works out the mean of the current month and the two previous ones.

This is copied across so that, for example, E3 holds '=(C2+D2+E2)/3'

We can now see that on average, sales are rising very nicely – indeed it looks like we are onto a significant upward trend. Are we?

Trends

The main interest in looking at a trend is what it might tell us about the future. What sort of sales can we expect next year? Will house prices rise, and by how much? How many passengers can we expect in 2015? If you can anticipate what is going to happen, you can increase profits, reduce costs, and avoid problems. How far can we look ahead with any degree of certainty? And how can we do it?

Spreadsheets help. We've seen how running averages can show trends – up to the present. The trick is to extend them. There are some very clever formulae which can take a series of numbers and project the trend from them – but they are too complex to be tackled here, and are in any case unnecessary. Graphing numbers and then extending the line can do just as good a job, and far more easily. But you have to do it sensibly.

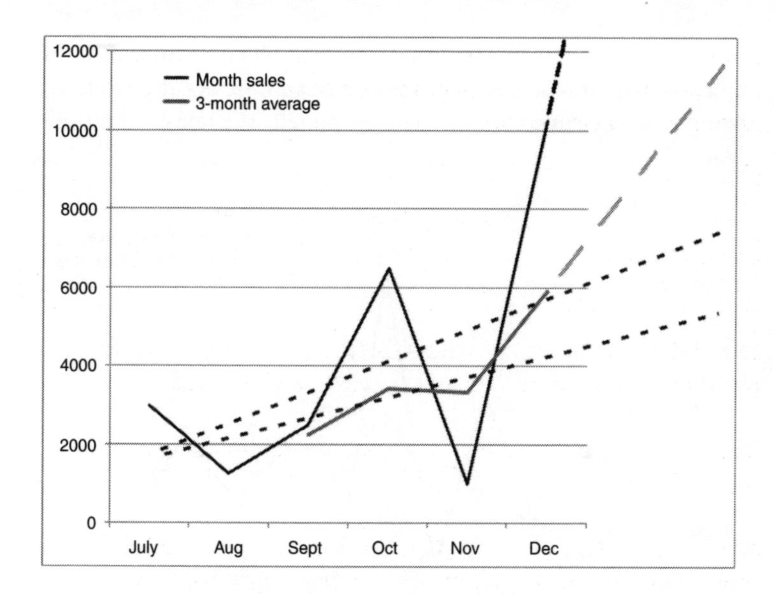

Look at the first graph. This was (largely) produced by the Chart facility in Excel. The two solid lines link the month sales and running average figures from that last table. The last section of each line is then extended as prediction of future sales. The dark broken line is clearly not to be trusted – that line has been so up and down already that you would be foolish to think that the last section really pointed the way ahead. The extension of the running averages line (in grey) is probably more realistic, but its last plotted point was partly based on the very high December figure, so caution is necessary. The two dotted lines overlying the graph represent more cautious interpretations of the trend – the future sales are more likely to fall in the area between these arms.

Let's take the next six month's sales and see how things panned out.

	July	Aug	Sept	Oct	Nov	Dec	Jan	Feb	Mar	Apr	May	Jun
Month sales	3000	1250	2500	6500	1000	10250	2	4500	3250	5000	2500	1500
Average			2250	3417	3333	5917	3751	4917	2584	4250	3583	3000

Ah! The early promise was not maintained. People don't buy many books in January, in any case, then after an improvement in the spring, sales began to tail off. The graph tells the tale even more clearly.

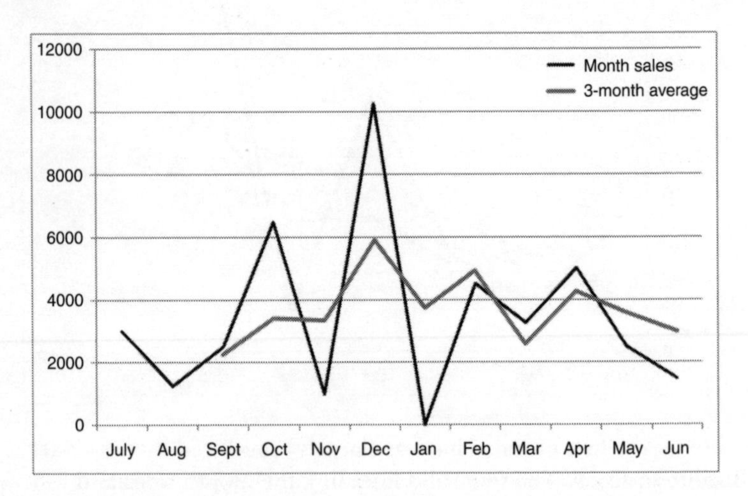

You can see here, compressed into a year, the story of most books, discs, fashion accessories and other consumer items. Sales rise, but after a while they start to fall as increasingly the sort of people who might buy one, already have one.

We find a different story with house prices. Over the 10 years to 2007, the average house tripled in price, from £66,313 to £183, 959.

1998	1999	2000	2001	2002	2003	2004	2005	2006	2007
66,313	74,638	81,628	92,533	115,940	133,903	152,464	157,387	172,065	183,959

House prices 1998 to 2007, figures from Nationwide

The rise was very smooth. Look at the graph. The dotted line here is the trendline, calculated by Excel. In 2007, if you had used that trend to predict house prices, you would have expected the average house to have reached about £240,000 by 2010.

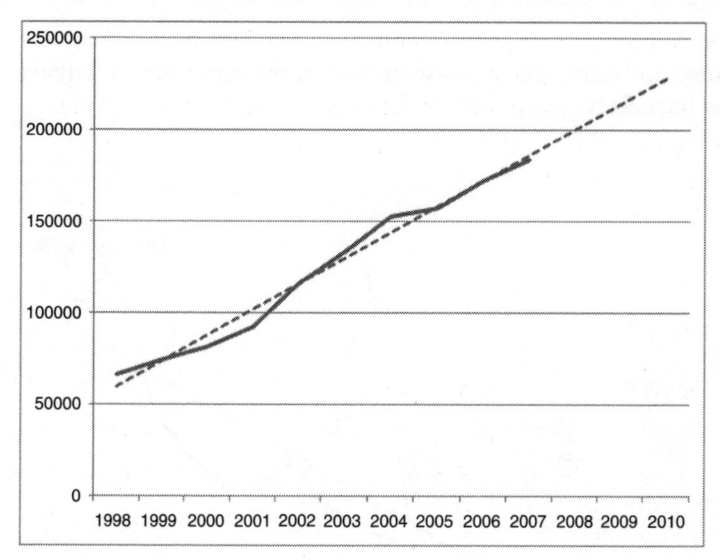

House prices 1998 to 2007, with the trend projected to 2010

But the reality was different – as we all know. The financial crash of autumn 2007 produced a drop in prices, and though they have picked up since, they are still some way off their high.

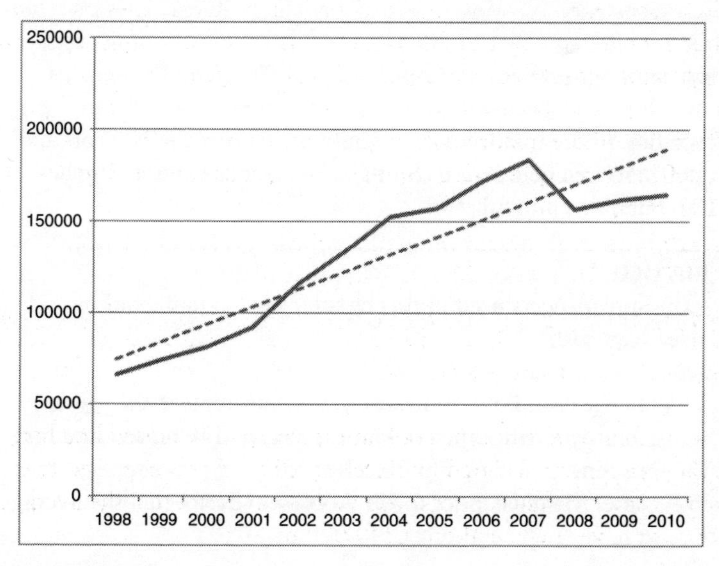

House prices 1998 to 2010

One more graph. If you take a longer view, you get a different trend line again, but you also see one of the limitations of relying on formulae.

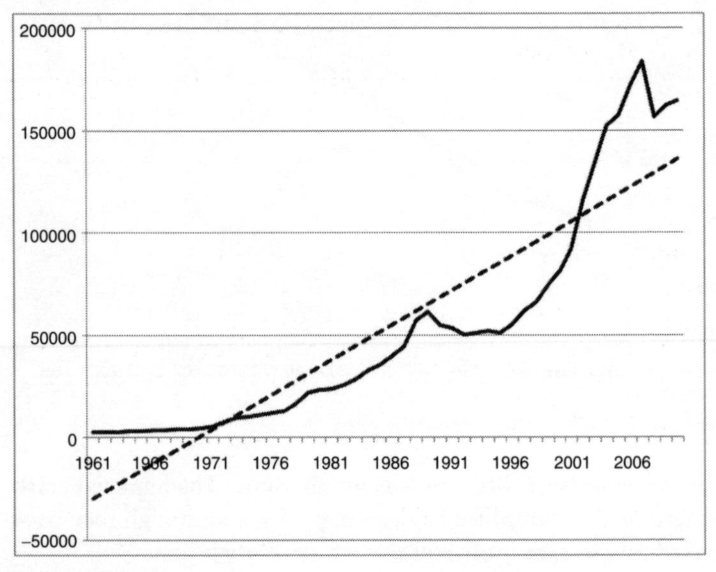

House prices 1961 to 2010

Here we've got 50 years of house prices. The trend line tells us that prices have been way over the trend these past 10 years – but according to the trend, houses should have been free in 1968 and better than free before that. Straight line trends cannot be relied on to tell a meaningful story.

Practice

The spreadsheets used in this chapter are available online (see page vii).

1 In the cash flows workbook, you can replace the random numbers with typed in values, or give different values for expense, salaries and premises costs, or the commission rate. Make changes to the Accurate sheet – the Rounded

sheet draws all its values from there. Explore the effect of any changes on the results.

2 On the Breakeven sheet, you can change the cost price and sale price, easily. If you can cut the unit cost to £3.00, what's the breakeven point with a sale price of £6.99, or £5.99?

7

Averages, ratios and proportions

In this chapter you will learn:

- ***Why and when these skills are useful***
- ***About different types of 'average'***
- ***How to calculate with ratios***
- ***About some ratios that are used in accountancy***
- ***About costing practices using proportions***

Why and when these skills are useful

Averages and ratios are routinely used as benchmarks or the basis for comparisons in many contexts in business, government and research. They are also routinely misused by politicians, advertisers and other people who are not to be trusted. Either way, it is as well to understand what averages and ratios can or cannot tell us, and even more fundamentally, what is meant by 'average'.

What do you mean by 'average'?

People use 'average' to mean 'the middle value' or 'the most common value', and in casual conversation this is fine. If you talk about 'the average bloke' this will be taken to mean 'most men' or more often 'most men like me', but your listeners won't expect this to be based on anything other than your gut feelings. On the other hand, if the 'average' is being put forward as an exact value, then you need to be clear about what sort of average it is. There are three ways of describing middling values:

- Mean
- Mode
- Median

MEAN

This tends to be what people are referring to when they say ‘average’. The mean is calculated by adding all the values in the set, then dividing the sum by the number of items in the set. For example:

Heidi jogged 6 miles on Monday, 4 on Tuesday, 5 on Wednesday, 6 on Thursday, 0 on Friday, 4 on Saturday and 10 on Sunday. What is the mean number of miles she jogged each day that week? The calculation is:

$$(6 + 4 + 5 + 7 + 0 + 3 + 10) \div 7 = 35 \div 7 = 5$$

We included Friday in the count of days, even though she didn’t run that day, because the question asked for the mean ‘each day that week’. If the question had been ‘What is the mean distance covered when she jogs?’ the calculation would have been:

$$(6 + 4 + 5 + 7 + 3 + 10) \div 6 = 35 \div 6 = 5.83$$

We have divided the total by the number of jogging days.

There’s an important point here. When data is collected, there may well be zero values, and these may or may not be included in the count. It all depends upon what is being measured and why. As so often with numbers, if there is confusion, it comes from the words around the numbers, and a lack of clarity about what the numbers are supposed to mean.

MODE

This is the most common value in a set. If you read in the papers that ‘the average woman is now size 12’, it means ‘more women are size 12 than any other size'. When people are writing about modes, or modal values, you will often see the text accompanied by a bar chart. The mode is the highest bar.

Our local shoe shop, Corns'R'Us, recorded the following sales of ladies shoes on Saturday:

Size	1	2	3	4	5	6	7	8	9
No. sold	4	12	36	42	25	15	10	7	2

The mode here is size 4 – more women bought size 4 shoes than any other size. You can see this clearly in the bar chart. Something else shows up in that chart. Although size 4 is the most common, there are a lot of size 3 and 5 shoes sold too. We've got what's called a 'normal distribution' here, with most values clustered around the middle, and tailing off above and below. Most measurements of populations tend to follow this pattern.

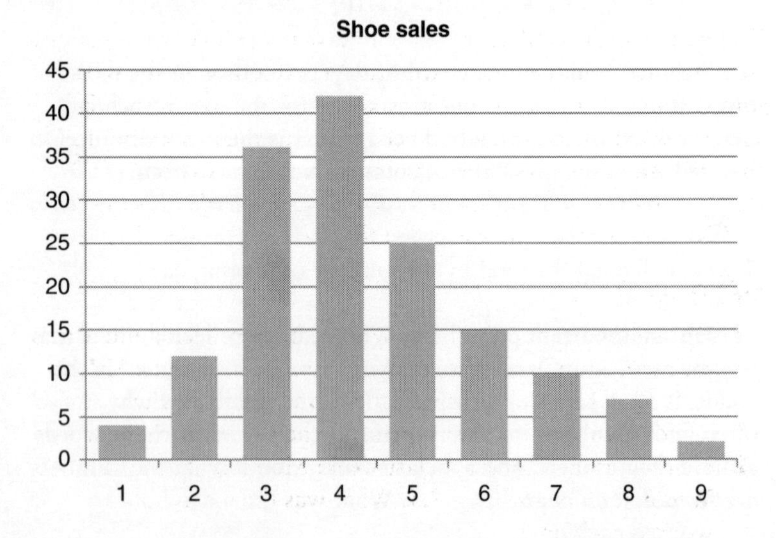

If you want to know the mean size of shoes sold on Saturday, you have to add up all the sizes, i.e. $4 \times 1 + 12 \times 2 + 36 \times 3 + 42 \times 4 + 25 \times 5 + 15 \times 6 + 10 \times 7 + 7 \times 8 + 2 \times 9 = 663$, then divide that by the total number of shoes sold (153) and that gives you 4.333.

And it's meaningless because they don't do shoes in that size. The mean is normally only useful where you have continuous variation, and decimals or fractions make sense. Where the things you are measuring go up in distinct steps, the mode is a more useful form of 'average'. The 'average' family does not have 1.8 children – no family has this specific amount.

MEDIAN

This is the middle number in the most literal sense. If you arrange all the values in a line, from lowest to highest, the median is the one in the middle of the line.

In McSpleen's Haggis Factory, where Mac works as a summer temp, there are 5 stuffers on £5,000 p.a., 3 mixers on £12,000 p.a., 2 cooks on £14,000 p.a., the foreman on £16,000 p.a. and Mrs McSpleen on £200,000 p.a.

To find the median value, we count how many people we've got (12) then starting from the lowest, work through halfway to the sixth lowest value, which is one of the mixers on £12,000. The median can be the most useful way to describe the 'average' in situations like this. The mean here is not very representative – it's over £25,000 (the boss' salary skews the result); neither is the mode (5 people earn £5,000), but there are more in total that earn well above that.

Practice 1

1. During the beer and skittles night at the pub, Heidi and Mac's 8 person team drank the following numbers of pints: 1, 1, 2, 5, 5, 5, 6, 15. Find the mean, mode and median values. Which do you think best represents their drinking habits?
2. Heidi worked these hours last week: Mon 9.5, Tue 10.0, Wed 8.75, Thur 9.0, Fri 8.5. What was the mean hours worked each day?
3. The pub holds quiz nights. The number of teams that turn up vary – over the past 10 nights there have been 7, 3, 9, 2, 5, 8, 6, 7, 4, 8. The landlord would like to know how

(Contd)

many tables he should set out, without overdoing it – empty tables take up space but it's a bother to have to set out extra as people come in. Find the mean, mode and median values, and from these recommend how many he should set out.

Ratios

A ratio is a way of expressing the relationship between two or more numbers. Ratios can be written like this: 10:3 or 5:2:1 with the numbers separated:by:colons. Or where there are only two numbers, the ratio can be written as a single value, e.g. 0.85, which is equivalent to 1:0.85. If it's not immediately obvious what the numbers stand for, there should be some explanatory text close by. For example, according to Epicurean.com, the perfect martini is in the ratio 3:1 (gin to vermouth), and in the Cosworth gearbox, first gear has a ratio of 2.95, which means the driveshaft turns 2.95 times to turn the axle once.

It's important to remember that ratios have no units. The 3:1 martini could be 3 capfuls of gin to 1 capful of vermouth, or 3 glasses of gin to 1 glass of vermouth, or any multiple of any unit as long as the ratio stays the same. So if you started with a bottle of gin (75 cl) you would need 1/3 of that (25 cl) of vermouth. (You'd also need a couple of litres of ice cubes, a 3 litre shaker and a strong bartender.)

The aspect ratio and where are the sides of my picture?

The aspect ratio of a TV, film or camera image is the ratio of its width to its height. The aspect ratio for analogue (i.e. what used to be, and still is in many places, standard) TV is 4:3, so an image might be 16 inches wide by 12 inches high. Widescreen, digital TV sets have a screen aspect ratio of 16:9. Modern movies are usually either 1.85:1 or 2.39:1 (we think they ran out of round numbers). 1.85:1 is fractionally wider than 16:9 (it's actually 16.65:9), so when such a movie is broadcast for digital TV only the very edges of the images are lost. 1.85:1 is far wider than 4:3 (it's 5.55:3) so when movie gunslingers face off on an analogue TV, all you can see is the tips of their guns poking in from either side. With cinemascope on analogue TV, nearly half the image is lost off the side if they broadcast it full height, which is why they often use the 'letterbox' display with image at almost half height.

CALCULATIONS USING RATIOS

Working with ratios is similar to working with fractions. In fact, you can express any (whole number) ratio as set of fractions – just add up all the numbers to get the denominator. For example, with 3:1 gin to vermouth, 3 + 1 = 4 and the recipe could be expressed as 3/4 gin and 1/4 vermouth.

Concrete is a mixture of cement, sand and gravel in the ratio 1:2:4. If I've got 18 bags of gravel, how much sand and cement will I need? To work this out you need to first find the multiple – what do you have to multiply 4 by to get 18. And to find that, you divide: 18 ÷ 4 = 4.5. Now we can multiply the other numbers in the ratio by this, and it gives us:

1 × 4.5 = 4.5 bags cement
2 × 4.5 = 9 bags sand

Sticking with the concrete – not in it – let's explore the ratio from the other end. The builders need 12 cubic metres of concrete to fill the foundations, and intend mixing it themselves. How much will they need of each? To solve this, we first need to find out what the ratio numbers add up to. (We are doing a fraction thing here.)

1 + 2 + 4 = 7

Then divide 12 by 7 to get the size of one part:
12 ÷ 7 = 1.714 cu metres (that's the cement)

Then multiply to find the other values:
2 × 1.714 = 3.428 cu metres (sand)
4 × 1.714 = 6.857 cu metres (gravel).

Practice 2

1 Mac's moped has a two-stroke engine which runs on a mixture of petrol and oil in the ratio 40:3. When he fills the tank with 5 litres of petrol, how much oil should he add?
2 Uncle Ebeneezer specified in his will that his estate was to be split between his three ex-wives in the ratio 5:11:2 – these being the number of years each put up with him. He left £2.7 million. How much did each ex get?

Ratios for assessing value

Ratios are much used in accountancy for assessing the health of a company.

The **liquidity ratio** is a measure of how well a business can find the cash it needs to meet its short-term debts. It is calculated by the simple formula:

$$\text{Liquidity ratio} = \frac{\text{current assets}}{\text{current liabilities}}$$

Current assets are cash or those things that should be turned into cash during the course of the next year's trading, i.e. stock, debts owed to the business, cash in the bank and in the till. Current liabilities are those debts which must be repaid during the next year's trading, principally outstanding invoices, and company tax. The ratio should be more than 1:1, and in most businesses it should be between 1.5:1 and 2:1.

The **quick assets ratio**, also called the acid test ratio, is a better guide to the ability of a business to survive a crisis. This uses only the most liquid assets and short-term liabilities – those things that can be turned into cash in a hurry:

$$\text{Quick assets ratio} = \frac{\text{cash} + \text{debtors} + \text{cashable deposits}}{\text{short-term current liabilities}}$$

If this is not above 1:1 then the business is in trouble, because it will not be able to pay its bills as they fall due.

Those accounting ratios that fall below 1:1 are expressed as percentages, i.e. 10% rather than 1:0.1 – it looks neater and is less likely to be misunderstood or mistyped. Here are a few of the commonly used ones.

The **gross margin ratio**, which looks at the profit on sales, without considering other overheads. To find it, divide the gross profit (sales – cost of sales) by sales:

$$\frac{\text{sales} - \text{cost of sales (cost of stock + commission + p \& p + other direct costs)}}{\text{sales}}$$

e.g. on £50,000 sales, stock cost £30,000, commission was £2,500, other direct costs £5,000, so you get:

$$\frac{£50,000 - £37,500}{£50,000} = \frac{£12,500}{£50,000} = 25\%$$

The **ROCE** (return on capital employed) ratio compares profits and assets – the basic question here is, 'Would the business be better off flogging everything it owns and putting the money in the bank to earn interest?'

The equation is simple: ROCE = net profit ÷ total assets

If a business had total assets (stock, vehicles, premises, cash, etc.) of £1,000,000 and made a net profit of £50,000, then its ROCE ratio would be 5%.

Practice 3

1 Offalori Trading's accounts show the following:
Stock £10,000
Debtors £5,000 (money owed to them)
Bank £1,500

(Contd)

Cash £6,000
Creditors £9,000(which they must pay)
Outstanding tax £16,000
What does the liquidity ratio tell us about the business? Use the quick asset ratio to see if they could survive a sudden call for cash.

2 As well as the stock, Offalori Trading has vehicles worth £25,000 and a warehouse worth £350,000. Last year they made a net profit of £40,000. What is their ROCE ratio?

Costing using proportions

Businesses have different types of costs:

- **Overheads**, such as rent, vehicles, administration staff and office expenses, which have to be paid for no matter how much the business sells.
- **The cost of the goods sold**, e.g. raw materials and labour in manufacturing, or suppliers' bills in retail, which vary with the quantities of goods sold.
- In manufacturing there are also **machine costs** which are very largely constant no matter how many hours a day they are in use – the energy they take when in use is part of the cost of goods.

If the business deals in two or more distinct products, that need different amounts of inputs and have different costs, then if it is to know the profitability of each type, it needs to find a way of apportioning the overheads and machine costs between them. Why does it need to know this? Because if it does not know, then it may find itself spending too much of its energies on the less profitable lines.

Dodger and Bodger the builders specialize in hand-crafted summerhouses and in loft conversions. The prices of both are worked out the same way: cost of materials + labour + 50%. They've increasingly been concentrating on loft conversions but don't seem to be making as much money. What's happening?

The problem is that loft conversions often need planning permission and always have to meet building regulations so a lot more time is spent on paperwork and meetings with council staff, and the cost of this is lost in the general overheads.

COST ACCOUNTING

In the 300 years or so since the industrial revolution and the beginnings of modern management, accountants and managers have devised several different approaches to analysing the costs in their businesses. Historically, most have tended to work along similar lines, taking the proportion of the labour costs and/or raw materials and other direct costs that have gone into each product line, and allocating the same proportion of the overheads and other costs to each line.

The method currently favoured by business is activity based costing, or ABC. This aims to identify the costs of all the activities in the business, and to establish how much each product uses each activity. This shows up which products have higher overheads, and management can then either seek ways to reduce its overhead, or increase prices to cover costs better.

The Acme Widget Co has two product lines: the basic widget, and the executive model, which has moving parts and a high-sheen metallic finish. The raw material costs are much the same, but each executive model takes twice as much labour and machine time. Acme produces 50,000 basic and 10,000 executive widgets a year. Under the traditional cost accounting methods, the overheads would be allocated in the proportion 50,000 to 10,000 × 2, or 5:2. So, 5/7 of the overheads would be borne by the 50,000 basic widgets, and 2/7 by the 10,000 executive widgets. An ACB accountant coming into Acme would see that the executive models have additional costs after manufacturing. Their moving parts need to be tested, and the finish checked for flaws; there is a higher reject rate, and a higher rate of returns for repair; and a disproportionate amount of the advertising budget is spent on them. If the firm has a realistic view of the total costs that go into a product, they can make sensible decisions about its future.

Though originally devised for manufacturing industry, the techniques can equally well be applied to financial institutions, media firms and other types of commerce. Here the main costs will generally be personnel, but the same activity-based analysis can be done.

The cost of costing

Apportioning costs is an inexact art, no matter what a cost accountant may claim. At some point there are always assumptions about how resources could otherwise be used and how profitable this might be. And costing does itself cost money, because staff time is spent on calculating and recording costs. When centre-based costing was introduced into the public sector in the 1990s, departments and sections within organizations became responsible for their own costs and were expected to price for staff or resources that were used by other departments. It was supposed to make everyone more cost-conscious and to make the organizations leaner and more efficient. In most cases it significantly increased the amount of time spent on administration, and reduced the proportion of money spent on front-line services.

Answers and explanations

Practice 1

1 The mean is (1 + 1 + 2 + 5 + 5 + 5 + 6 + 15)/8 = 40/8 = 5.
The mode and median were also 5, so in this case however you like to work it, the average was 5.

2 The mean daily hours of work was
(9.5 + 10.0 + 8.75 + 9.0 + 8.5)/5 = 45.75/5 = 9.15 = 9 hrs 9 mins.

3 The mean number of teams was
(7 + 3 + 9 + 2 + 5 + 8 + 6 + 7 + 4 + 8)/10 = 59/10 = 5.9.
This had two modes – 7 and 8.
The median number was 7.
The landlord should set out 7 tables.

Practice 2

1. To work this out, we divide 5 litres by 40 = 5000 ÷ 40 ml = 125 ml, then multiply that by 3: 125 × 3 = 375 ml.
2. The ratio here adds up to 5 + 11 + 2 = 18. Divide £2.7 million by 18 to get the size of each share = £150,000, then multiply by the ratio numbers. The first ex-wife gets £750,000, the second gets £1,650,000 and the third gets £300,000.

Practice 3

1. Offalori Trading's Liquidity ratio was

$$\frac{\text{£10,000 (stock)} + \text{£5,000 (debtors)} + \text{£1,500 (bank)} + \text{£6,000 (cash)} = \text{£22,500}}{\text{£9,000 (creditors)} + \text{£16,000 (tax)} = \text{£25,000}}$$

= 1:0.9

Their quick assets ratio was:

$$\frac{\text{£5,000 (debtors)} + \text{£1,500 (bank)} + \text{£6,000 (cash)} = \text{£12,500}}{\text{£9,000 (creditors)}}$$

= 1.39

While they can survive in the short term, they do not have the wherewithal to meet their liabilities.

2. Offalori Trading's total assets are:
£22,500 (current assets) + £25,000 (vehicles) + £350,000 (warehouse) = £397,500.
Their ROCE was £40,000 ÷ £397,500 = 10%.

8

Finding the unknown: *x* and *y*

In this chapter you will learn:

- ***Why and when these skills are useful***
- ***Basic rules of algebra***
- ***Using algebra in word problems***
- ***Translating English into algebra***

Why and when these skills are useful

Simple algebra – which is all we are trying to cover here – is a technique for finding the unknown. Perhaps we'd better qualify that. It is a technique for finding missing numbers, in situations which can be expressed as equations.

Algebra can be a way to get to the numerical heart of a problem. By applying its rules and techniques you will be able either to solve the problem or to identify that there are too many unknowns for a solution to be possible – and that can sometimes be as useful as finding a solution.

Basic rules of algebra

Let's start with a practical example, and we'll then draw the theory from concrete.

Heidi: ‘I just bought 2 apples. Do you know how much they were each?’

Mac: ‘Och lassie, I’ve nae clue.’

The price of apples will have to stay unknown because there is nothing to work with. Let’s try it again. (This time without the accent, ed.)

Heidi: ‘I just bought 2 apples and only got 20p change out of a pound. How much were they each?’

Mac (quick as a sloth): ‘40p?’

The workings behind the answer were:

1	2 apples cost £1.00 – 20p	$2a = 100 - 80$
2	2 apples cost 80p	$2a = 80$
3	1 apple cost 40p	$1a = 40$

At step 1 we summarize the situation and turn it into an equation, using a letter to stand for the unknown value.

At step 2, we simplify the known numbers, which tells us that 2 apples cost 80p.

At step 3, we divide both sides by 2 to find the cost of one apple.

We now have the single unknown (1a) on one side, and its value (40) on the other. Solved!

And there you have the key moves in algebra:

- You must be able to express the situation as an equation.
- Numerical values should be simplified, by working out sums, as far as possible.
- Whatever you do to one side of the equation, you must do to the other to keep it in balance.

Let's walk through another example. We're back at the shops and have bought a magazine for £1.50, and three bars of chocolate. The assistant gives us £1.10 change from a fiver. How much does a bar of chocolate cost? Apply the rules:

1. Express the situation as an equation: we'll write what we got (magazine, chocolate and change) on one side, and what we gave (£5) on the other.
   £1.50 (mag) + 3 bars + £1.10 (change) = £5
2. Reduce this to numbers and a letter to stand for the unknown
   $1.50 + 3b + 1.10 = 5$
3. Simplify the sums
   $1.50 + 1.10 = 2.60$
   $3b + 2.60 = 5$
4. I want to get the unknown on one side of the equation and the numbers on the other. We can perform any operation on an equation as long as we do the same to the other side. Subtracting 2.60 from both sides will leave 3b alone on the left.
   $3b + 2.60 - 2.60 = 5 - 2.60$
   $3b = 2.40$
   Then divide by 3:
   $b = 2.40 \div 3$
   $b = 80$

Translate back into English: one bar of chocolate costs 80p.

You can always find the solution if there is only one unknown value.

Practice 1

1. 3 pints and a packet of crisps cost £8.00. The crisps cost 50p. How much is a pint?
2. Heidi is going to make marmalade and has been shopping on her bicycle. She has 16 oranges and 2 kilos of sugar in the bag on the left of her handlebars and 6 oranges and 4 kilos of sugar in the bag on the right. The two are perfectly balanced. How much does an orange weigh?

3 x is the unknown. Find the different values for x in these three equations:

a $2x = 14$

b $x - 6 = 9$

c $4x + 2 = 30$

What if there is more than one unknown?

Ah, now that depends. Let's go to the pub and have a think about it.

Mac buys the first round: 2 pints and 3 martinis cost £14. How much is a pint? The answer is, 'We can't work it out.' There simply isn't enough information.

Heidi buys the next (smaller) round: 1 pint and 2 martinis cost £8.50.

Now we can do it. We can set up equations to describe both situations, then work them side by side to reach a solution. We're into *simultaneous equations*, which you may vaguely remember from school days.

$$2p + 3m = 14 \qquad (1)$$
$$1p + 2m = 8.50 \qquad (2)$$

The trick is to take one of unknowns out of the equations, because we can solve it if there is only one. To do this we need to have the same number of either pints or martinis.

Multiply equation (2) by 2 – and note that we multiply everything on both sides

$$2p + 4m = 17 \qquad (2)$$

Look at the two equations. What is the difference between them? They both now have 2 pints, but (2) has one extra martini, and costs £3.00 more. You can work this out by subtracting the values in (1) from their equivalent in (2)

$$2p + 4m = 17 \qquad (2)$$
$$2p + 3m = 14 \qquad -(1)$$

So $1m = 3$

You can now go back to either of the original equations, and replace each martini with £3.00

$$2p + 3 \times 3 = 14$$
$$2p + 9 = 14$$

Subtract 9 from either side:

$$2p = 14 - 9 = 5$$
$$p = 2.50$$

From English to algebra

To convert a problem expressed in words into algebra, you need to get to its essentials. Look for the things you don't know and assign a letter to each of them. Look for the numbers, and see how they relate to the unknowns. And look for the possible equations. For example: there's a hole in Mac's sporran, just large enough for 5p and 1p pieces to slip through. On his way home one evening, he loses 11 coins, to the value of 27p. How many did he lose of each type.

What don't we know here? We don't know how many 5p coins or 1p coins. We'll call these s (for silver) and b (for bronze).

What do we know? We know there were 11 coins in total, so:

$$s + b = 11 \qquad (1)$$

We also know that the total value was 27p, so by bringing in the value of each type of coin we get:

$$5s + 1b = 27 \qquad (2)$$

And we've now got a pair of simultaneous equations that we can solve in the usual way. There's 1b in each equation and (1) is smaller than (2) so we'll subtract that.

$$\begin{aligned} 5s + 1b &= 27 && (2) \\ s + b &= 11 && -(1) \\ 4s &= 16 \end{aligned}$$

Divide each side by 4:

$$s = 4$$

Substitute 4 for 2 in (1):

$$4 + b = 11$$

Subtract 4 from each side

$$b = 7$$

Mac lost four 5p coins and seven pennies.

Where do *x* and *y* come into it?

Traditionally, the unknowns are represented by x and y, or a and b, and written in italics. But this is optional. Using the initial letters of the things that they stand for helps to remind you that there is a real world behind the equations.

Practice 2

1 Mac has offered to help with the marmalade making. He was asked to buy 12 oranges and 3 lemons, but misheard and bought 3 oranges and 2 lemons. They cost 60p. He then

(Contd)

had to go back and get the other 9 oranges and 1 lemon, which cost £1.05. Heidi said, 'Oranges have gone up. They were only 6p each last year.' How much did Mac pay for them?

2 All 100 tickets for the Southampton Ukulele Jam's annual concert have been sold for a total of £680. The tickets cost £8 full price or £5 concessions. How many people paid full price?

Answers and explanations

Practice 1

1 3 pints + 1 crisps = 800
1 crisps = 50
3 p = 750
p = 250. 1 pint costs £2.50

2 16 oranges + 2 kilos = 6 oranges + 4 kilos
Subtract 6 oranges from each side
10 oranges + 2 = 4
Subtract 2 (kilos) from each side
10 oranges = 2 kilos = 2000 gms
1 orange = 200 gms

3 **a** $2x = 14$ Divide by 2
$x = 7$

b $x - 6 = 9$ Add 6 to each side
$x = 15$

c $4x + 2 = 30$ Subtract 2 from each side
$4x = 28$ Divide by 4
$x = 7$

Practice 2

1 We can get two equations from this:
3 oranges + 2 lemons = 60p (1)
9 oranges + 1 lemon = 105p (2)

If we double (2) we will have the same number of lemons in each. The bigger numbers will be in (2) so we'll subtract (1) from it:

18 oranges + 2 lemon = 210 (2)
3 oranges + 2 lemons = 60 – (1)
15 oranges = 150
1 orange = 10p

2 If we use f for full price and c for concessions, we can get these two equations:

$f + c = 100$ (number sold) (1)
$8f + 5c = 680$ (ticket money) (2)

Multiply (1) by 5, then subtract this from (2).

$8f + 5c = 680$ (2)
$5f + 5c = 500$ (1)
$3f = 180$
$f = 60$

60 tickets sold at full price.

9

Probability and risk

In this chapter you will learn:

- ***Why and when these skills are useful***
- ***How probability can be expressed***
- ***How chances can combine***
- ***And an odds client day at the races***
- ***Risk (absolute and relative and writing about risk)***

Why and when these skills are useful

A lot of people do not understand probability – if they did, they would not donate so much of their money to the Lottery. And in business and finance, people do not always assess risks properly – if they did we would not have had the recent bank crisis. Some aspects of probability are counter-intuitive – they seem to run against what you might expect – but that does depend on how you look at them. In the commercial world, confusion can come from over-calculating, so that the guesses on which assessments are based are lost in a welter of apparently accurate figures.

Let's start looking at probability from first principles.

What's the chance?

Here's the situation. You work in a branch office, which the boss always visits once a week and only once a week – but on no particular day. That means:

- On any given day, there's a 1 in 5 chance that he will visit.
- On any given day, there are 4 chances in 5 that he won't visit.

The odds shorten as the days pass:

- If he doesn't visit on Monday, there's a 1 in 4 chance that he will turn up on Tuesday (or later).
- If he doesn't visit on Tuesday, there's a 1 in 3 chance that he will turn up on Wednesday (or later).
- And so on until the end of the day on Thursday. If you haven't seen him by then, he is certain to come in on Friday.

But once he does visit, there is no chance that he will come again.

Expressing probabilities

You can express probability, and calculate with chances, in a number of ways. 1 in 5 can equally well be written as:

1:5 20% 1/5 0.2

Use whichever form will enable you and other people to understand the situation most clearly, and be ready to switch between them when you are doing calculations. In practice, fractions or percentages are used most of the time.

We can draw two key principles from the boss's visits:

- Events either happen or they don't happen. If an event is absolutely certain to happen, its probability is 1. If it is impossible, the probability is zero. If it may or may not happen, the probabilities of the outcomes must add up to 1. $1/5 + 4/5 = 1$.

'Ah, but what about fireworks?' you ask. 'They can either get lit and go bang, or not be lit or get lit and not go off.' And the answer is: 'Not every event is an either-or case. There can be multiple outcomes. From that box of Blasto crackers that Heidi bought last November, 7 went bang, 2 fizzled out after lighting, and 1 was left in the box. So the chances were $7/10 + 2/10 + 1/10 = 1$.

- Probabilities may change over time or in response to other events. Because the boss visited *once and only once* a week, the likelihood of him visiting increased every day until he appeared. If he changed his routine so that he visited *on average* once a week – but sometimes not at all and sometimes twice or more – then the probability of him turning up on any given day would always be 1 in 5.

COMBINING CHANCE

It might help if you had a few coins to toss while you are reading this next bit.

If you have just tossed nine tails in a row, what is the probability that you will get a head next time? You might think that after a string of nine tails, it was almost certain to fall heads, but that's not so. Each time you toss a coin, there is a fifty-fifty chance of heads or tails. This is just as true the tenth time as it is the first time. 'But, but, but...' you are saying. Of course it is also true that it is very unlikely that anyone will toss 10 tails in succession, but you've just had 9 tails – and that was also unlikely – and at this point 10 tails is just as likely as 9 tails and a head.

Each toss is a separate event, so its odds remain the same, but if you combine the events by simultaneous flipping or by treating the sequence as a whole, then the probabilities change.

If two of you toss coins at the same time, or you toss two one after the other, the outcomes can be:

HT HH TH TT

You will get at least one head 3 times out of 4, while the chance of zero heads (TT) is 1 in 4.

Try it with 3 coins, and can get:

HHH HHT HTH HTT THH THT TTH TTT

Now the chance of zero heads is 1 in 8. That is the same of course for any given outcome – HTH or TTH equally have a 1 in 8 probability. Each time you add another coin, the chance of any given outcome is reduced by a half.

T	TT	TTT	TTTT	TTTTT	TTTTTT
$\frac{1}{2}$	$\frac{1}{4}$	$\frac{1}{8}$	$\frac{1}{16}$	$\frac{1}{32}$	$\frac{1}{64}$

So what are the implications of this? If you were betting separately on the outcome of each toss, and hitting a string of tails, you would need to increase your bet each time if you wanted to win money overall. If you bet £1 on the first toss, you would have to bet £2 the second time to win back what you have lost and come out ahead, then £4 the third time, £8 next, and so on. By the eighth toss you would have to bet £256 – and even then would only be £1 ahead if you won. But whoever you were playing against would have walked off with their winnings in their pocket long before then.

The same principle applies in any situation where you have a sequence of events, but the outcome of each is independent of the rest. Which brings us back to the lottery. If you always play the same numbers, the fact that they have not come up once in the past 10 years does not make it any more likely that they will come up next week. Their turn will not, necessarily, come.

DICE

Probabilities can be combined in different ways. Take dice, for example, where the combined total is more important than what's on each individual dice. If you are a regular player of backgammon, Monopoly or any other game that uses 2 dice, you will know that double 6 is rare, but middling scores come up all the time.

	⚀	⚁	⚂	⚃	⚄	⚅
⚀	2	3	4	5	6	7
⚁	3	4	5	6	7	8
⚂	4	5	6	7	8	9
⚃	5	6	7	8	9	10
⚄	6	7	8	9	10	11
⚅	7	8	9	10	11	12

Total	Probability	Total	Probability
2	1/36	8	5/36
3	2/36 = 1/18	9	4/36 = 1/9
4	3/36 = 1/12	10	3/36 = 1/12
5	4/36 = 1/9	11	2/36 = 1/18
6	5/36	12	1/36
7	6/36 = 1/6		

There are 6 different ways to score 7, but if only the total matters then the probabilities can be combined by addition (1/36 + 1/36 + 1/36...). Likewise, if you needed at least 10 to win the game, the chance of this are: 3/36 + 2/36 + 1/36 = 6/36 = 1 in 6.

Insight

If you are a games player, and want to know the odds, there's a print-ready version of the two dice crib sheet available for download.

What is true for dice is true for any situation where there are two or more sets of outcomes, but it is the overall impact that matters. For example, Giza Trading has two stores. Store 1 estimates that there is an equal probability of them making a loss or profit of £10,000 or of breaking even. Store 2 predicts four possible outcomes, all equally likely. The numbers are shown in the next table. What is the probability of Giza Trading showing a profit overall?

	−10,000	0	+ 10,000
−10,000	−20,000	−10,000	0
0	−10,000	0	+ 10,000
+5,000	−5,000	+5,000	+ 15,000
+10,000	0	+10,000	+ 20,000

There are 12 possible outcomes overall, of which 5 show a profit.

For simplicity all outcomes have been taken as equally likely. In practice, once you get away from the dice board, different outcomes have different probabilities and that complicates the overall results. We'll get back to this.

CARDS

A pack of cards provides examples of other ways in which probabilities combine. When you take a card out of a full pack there is a 1/52 probability of selecting any specific card. If you don't pick it first time, the chance of picking it next time go down to 1/51 because there are only 51 cards left – and if you do pick it, the chance of picking another specific card are likewise 1/51 for the same reason. For the third pick, the chances are 1/50 and so on down.

The probability of picking the ace, king and queen of spades – in that order – are:

$$\frac{1}{52} \times \frac{1}{51} \times \frac{1}{50} = \frac{1}{132{,}000}$$

Which is why you don't see it very often.

The odds are improved if you don't care which order they are picked in. You could have AKQ, AQK, KAQ, KQA, QAK, QKA – that's 6 combinations, so the odds are now 6/132,000 – 1/22,100. If you don't care which suit, then the odds improve again by a factor of 4 to just over 1/5,000.

With 13 cards in each suit, what are the chances of picking 2 cards of the same value? Let's work it for aces.

The odds of picking an ace from a full pack is $4/52 = 1/13$
The odds of picking a second ace is $3/51 = 1/17$
So the odds of picking both are $1/13 \times 1/17 = 1/221$

There are 13 different values, so if any are equally acceptable, the odds of pickling a pair are $13/221 = 1/17$.

In most card games, players are dealt a hand which they may be able to improve by discarding or exchanging cards one way or another. They may be selecting 3 from an initial 5 or 4 from 6, and the odds of getting certain combinations are quite complex to work out. So much so, that we are not even going to try to do it here.

The real-world lesson to draw from cards is this: if the success of any project depends upon a series of separate factors, each of which may produce a range of results of which only one is acceptable, then the more factors there are, the less the likelihood that the project will succeed – and the odds against success will rise rapidly with each additional factor. If a number of alternative results are acceptable for each factor, then the odds of success rise sharply. On a complex project you must therefore either build in flexibility at every stage, or have stringent quality control and tight project management.

Play smarter – know the odds

If you are a poker player, or are simply interested, there is a very good article on the odds of different types of 5 card poker hands at Wikipedia. Look for 'Poker probability'.

The principles that apply to card probabilities also apply in any situations where each event or selection changes the odds for the next event. You might, for example, have a pool of 20 people, with varying talents, from which you need to assemble a team of 4 for a special project. How many ways can you do this? Don't try to answer that – it was a purely hypothetical question, and you do not have the data to solve it.

THE NATIONAL LOTTERY

The chance of buying a jackpot-winning ticket in the National Lottery is fractionally under 1 in 14 million. Of course, that means your chances are 1 in 14 million better than if you didn't buy one at all, but... Let's look at it another way. If you bought 10 tickets for each draw, that's 1,000 tickets a year, then the chance of winning a jackpot is 1,400. Keep doing it until AD3410 and you should have had one by then.

Small wins are much more likely, but not really worth having. The odds of matching 3 balls are only 1:55, but that only pays out £10, and the odds of matching 4 balls are just over 1 in 1,000, and that pays around £100. By combining these two, you see that on average for every £1,000 regular players spend, they will win 20 lots of £10 and one of £100, or £300 in total.

A DAY AT THE RACES

In horse racing and other forms of gambling based on sports or people's achievements (X-factor, elections, etc.), the odds are not based on mathematical probabilities, but on judgement. That is not to say that there is no maths involved – on the contrary, the bookmakers will have the numbers very carefully calculated. But the reason why the odds on Laughing Boy and Slipper Fetcher are both 10/1 while those on the Walthamstow Whizzer are evens, is that the bookies think the Whizzer is very likely to win. By offering longer odds on outsiders, they encourage people to bet on them, in the knowledge that this money will normally go into the pot from which they pay out on the winner, and leave some behind as profit.

What's the odds?

Bookies have their own special way of expressing odds. What do they mean? Some are more obvious than others.

Odds	What you win if your chosen horse/dog/team/X-factor contestant wins
10/1	£10 for every £1 you have bet, and you get your stake money back.
7/2	£7 for every £2 bet
Evens	£1 for every £1 bet
4/1 on	£1 for every £4 bet – this one is expected to win!

In horse racing, there are three ways you can bet on a horse:

Win

Place, i.e. to come in the first two (or three or four – it depends on the number of runners). If you win, you will get 1/4 (sometimes less) of the full odds.

Each way – win or place, which combines the two. If the horse does win, you will get the full odds, plus 1/4 for the place, and all of your stake back. If it is placed, you get 1/4 of the odds, but lose the 'win' half of your stake.

Why the odds change

If you had a 10 horse race, where all 10 horses where completely unknown, then the bookmakers might start with all of them on odds of 9 to 1 (10 to 1 wouldn't leave a profit margin). In the time between the odds being first given and the start of the race, as money comes in, if more of it goes on one horse than another, the bookies will shorten the odds on it – to 8 to 1 or less – partly to reduce what they will have to pay out, partly to encourage betting on other horses.

Practice 1

1 The area sales manager and the regional director both make a point of visiting every branch office once (and only once) a fortnight. If they haven't been already, what is the chance of one or other of them turning up on the first Wednesday?

What is the chance of them both turning up that day? The office is only open Monday to Friday.

2 Mac and Heidi are playing Backgammon. Mac needs to throw 9 or more to win the game. Heidi bets 5 smarties against his last 3 that he won't make it. Should Mac take the bet?

3 Slapdash & Hope Electronics sell a hi-fi system in which, in the first year, 1 in 10 of the amplifiers fail, as do 1 in 5 of the tuners and 1 in 20 of the speakers. What is the chance of reaching the end of the year with a system that works?

4 If Likely Loser is 2/1 odds-on favourite to win at Cheltenham, should I put a bigger or smaller bet on Willy Win, who is currently trading at 7–2?

Risk

Risk, as in the sense of risk assessment and risk management, is where gambling meets the business and organizational world. Risk is inescapable. You cannot run a business or organization without running risks, but for that matter you cannot run – or even walk – down to the shops without incurring risks. You could be there, walking carefully along the pavement, when a car swerves out of control, mounts the pavement and runs you down. It happens – indeed, it happens several times a year in the UK alone – but compared to the number of car journeys that are made entirely within the kerbs, and the number of foot journeys that are made without incident, the chance of it happening to you are slight. (We don't have the exact figures, but in the US, pedestrian fatal accidents account for 1.6 deaths per 100,000 people – and this includes misjudged crossings of busy roads, late night drunks and other riskier activities.) The point is, some events may be catastrophic if they occur but are very unlikely, and must be ignored because trying to avoid them would place impossible limitations on your life or your business.

There are two dimensions to risk:

- The probability that the event will occur.
- The consequences if it does.

People who write about risk management (or any other type of management) do like quadrant diagrams, so let's have one here. The horizontal axis represents the consequences, and the vertical axis the probability. Risks which fall in the bottom left (low probability and low consequences) can be ignored; risks which fall in the top right (high probability and high consequences) must be tackled as soon as they are identified. The action line – beyond which you take notice of risks – runs along the top-left/bottom-right diagonal. The trick is to work out where risks fit into the grid.

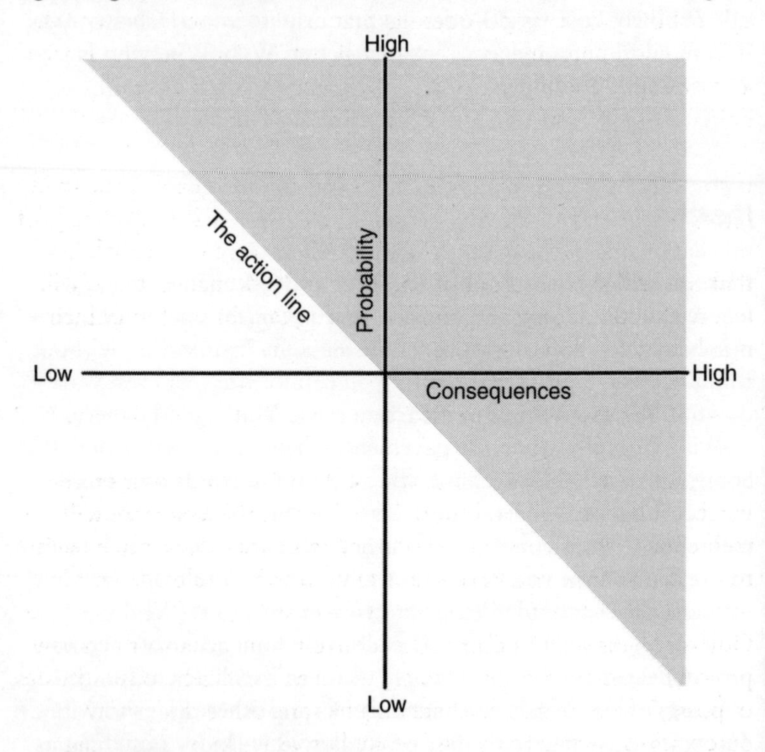

ASSESSING PROBABILITIES

As with gambling, in risk assessment the probabilities may be based on mathematical facts or verifiable numbers, or they may be based on judgement. It is important to recognize the difference between the two and to keep them separate. Unfortunately, some people don't. One method of assessing risks is to ask the people involved to rate them on a scale of 1 to 10 (or whatever range), then average them out, or otherwise calculate them. This approach might be used for assessing the likely profitability of several alternative projects, and result in one being rated 7.5, another 8.2 and the third 8.4. Now, as long as all of the assessors take the same view of risk and have applied the same criteria to each, this will give a usable rank order. But it is very hard to standardize judgement, and if different people are involved in the assessments, even the most sophisticated calculations won't produce results that can be properly compared. If Tom, Dick and Jane assess project A, while Tom, Sue and Harry assess project B and Sue, Ali and Dick assess project C, this is unlikely to give you a way of comparing A, B and C.

If you can, by whatever means, put a reasonably reliable probability figure on the risk – so that you can say, 'there is a 1 in 20 chance of this happening in the next 3 years', or 'there's a 25% probability that this will show a profit of £5,000 and 10% chance that it will lose £10,000', or anything similar – then you can start to crunch numbers when deciding what to do.

ASSESSING CONSEQUENCES

Some risks are readily quantifiable. If the refrigerator unit on the van breaks down half way to the fair, so that the ice cream will melt either before you can get it to the fair to sell it, or get it back to the depot, then you have lost a van's worth of sales.

Others are less so. The firm about to invest a huge amount in a new project has only estimates of costs, and even more nebulous estimates of possible sales to go on. There may be spreadsheet that show outcomes calculated to the last penny but, as we know, anything based on guesses will at best give only a range of possible result.

Some are beyond counting. If there is a fire in the office which destroys the computers and the back-up disks with the firm's accounting data, the firm could fold while it is trying to sort out its finances.

Health and Safety risk assessment

This is almost entirely about using judgement to identify hazards and taking reasonable steps to protect the workforce and members of the public. Numbers don't come into it, except in relation to the cost of protective measures – and whether these are implemented or not is a matter of judgement. We're not covering it here.

CROSSING THE ACTION LINE

The simplest decisions to take are where the risk and its probability can both be quantified. If you know that the failure of an essential part of the supply line would cost the business £50,000, and a failure is likely to happen once every 10 years, then it is clearly worth spending £5,000 (or something over that) a year to prevent it, or to insure against it.

If the consequences could break the business, then the risk must be avoided or insured against, if at all possible. An investment so risky that its failure could push the business into bankruptcy should never be undertaken – unless bankruptcy is already looming close and this is one last desperate throw. A high-end back-up system with secure off-site storage may be expensive, but could save the business if theft, fire or virus attack destroys its accounts data.

Where the consequences of any individual event are small, but the probability of occurrence is high, the cumulative costs should prompt action to reduce the risks.

Does management understand the risks?

The collapse of Barings Bank in 1995 and Lehman Brothers in 2008 were both ultimately for the same reason. The senior management did not understand the complex financial instruments that were in use, or the risks that junior employees were taking. In effect, they allowed juniors to 'bet the business' – and they lost.

Practice 2

1. KeepUSafe Inc have offered to insure on my premises for £3,000 p.a., covering everything except impact by meteor and falling aircraft. InDirect Line will cover absolutely everything for £3,500. Which one should I choose?
2. At the McScrooge Haggis Factory, they save 50p per haggis by sourcing their raw materials from Offalorri Trading and only pre-cooking the haggises enough to make them congeal. As a result, 1 in every 100 haggis sold results in sickness, and 1 in 5 of these cases are blamed on McScrooge. The firm has found that a letter of apology, plus a complimentary box of haggises and a bottle of whisky (total value £25) will satisfy the customer 49 times out of 50. The 50th will need an out of court settlement of £5,000. Is this a viable business model? Would it be cheaper to make haggises which were safe to eat?

Answers and explanations

Practice 1

1. By Wednesday, there are only 8 days left of the fortnight, so for each manager there is a 1 in 8 chance of a visit. The chances combine thus.
   One or other manager visiting: $1/8 + 1/8 = 2/8 = 1/4$
   Both managers visiting: $1/8 \times 1/8 = 1/64$
2. There are 4 ways to throw 9, 3 to get 10, 2 to get 11 and 1 to get 12. So the odds of throwing 9 or more are 10/36 that's almost 1 to 4. Heidi is offering 5 to 3 that Mac won't make it. On this outside chance, he needs bigger odds to risk his last smarties.
3. The chance of a failure in a Slapdash & Hope Electronics hi-fi system can be found approximately by adding up the component failure rates:
   $1/10 + 1/5 + 1/20 = 2/20 + 4/20 + 1/20 = 7/20$
   Strictly speaking, it is slightly lower than this as there will be some overlap, with two or more components failing in the same system.

4 A £2 bet would win £1 on Likely Loser, or £7 on Willy Win – but that is far less likely to happen. Do you feel lucky?

Practice 2

1 The only extra cover you get from InDirect Line is against meteors and falling aircraft. Unless the premises are under the flightpath of an airport used by dodgy airlines, the probabilities are so remote the extra £500 is not good value.

2 One way to assess this is to add the cost of placating the customers into the production costs of each haggis.

1 in 500 haggises costs £25 = 5p per haggis
1 in 25,000 haggis costs £5000 = 20p per haggis

The total additional cost is 25p, which is still less than the 50p they save. In the short term this is a viable business model. In the longer term it isn't, because once the food safety inspectors or the newspapers get wind of what they are doing, the business will be closed down.

10

Show me my money: personal finance

In this chapter you will learn:

- ***Why and when these skills are useful***
- ***Credit cards: APR, AER rates***
- ***Mortgages terminology: fixed, variable, interest-only, index-linked***
- ***Buying and leasing a car (amortization)***
- ***Simple and compound interest***
- ***Calculating tax***
- ***The life sum: when can I retire?***

Why and when these skills are useful

Most personal finance questions are based on interest – how much it costs to borrow, and how much it can earn if you are saving. Whether you are borrowing or saving, failure to understand interest rates can be an expensive shortcoming – and they are not difficult to understand, though banks, credit card companies and building societies may try to throw dust in your eyes. In this chapter we will try to show you how to get the best value for your money.

SIMPLE AND COMPOUND INTEREST

First you need to understand that there are different ways of calculating interest, and there are two key aspects to this:

- Simple and compound interest
- How often interested is calculated and paid

Simple interest is calculated on the initial capital – the amount invested or borrowed – throughout the time of the loan. It is typically paid, once a month or once a year, into a different account. Under this system, £100 invested at 6% p.a. will earn you £6 a year, every year. Some investments pay out on this basis. They are usually described as income bonds, or similar, and are typically for a fixed period of 2 to 5 years, paying interest each month.

With *compound interest*, the interest is added to the capital, so that in the next period the interest will be paid on the existing capital plus the last year's interest. £100 invested at 6% p.a. will earn £6 in the first year, but in the next year it will earn 6% of £106 = £6.36. Most savings accounts work on this basis, as do all loans and mortgages – no one, except Mum or Dad, will lend you money on simple interest.

How long will it take to double my money?

Compound interest makes your money grow. But how fast? How long would it take to double your money – assuming that you didn't add to the capital? The answer is not as long as you might at first think, but probably longer than you really wanted to wait. For instance, 5% pays out £5 a year on £100, you might think it would take 20 years to pay a total £100. Nope, the interest on the interest means that you'd double your money in just over 14 years. At 7% – the best rate on offer at the time of writing – it would double in 10 years. At 3% – the best *safe* rate on offer at present – it would take 23 years.

How long will it take for my debts to double

If you service your debts – just paying the interest as it becomes due, and not paying off any of the capital – then a debt will stay the same for ever. If you fail to pay the interest, then it is added to the debt, which increases the interest due, which makes it more difficult to pay off. Suppose you had a mad fling and maxed out your credit card, but then lost your job. At the moment, credit card companies are charging around 17% interest, and at this rate the debt would double in about 4 years 6 months. (Actually, as the company would be charging £20 a month non-payment penalty, it would double in 3 years 8 months.)

FREQUENCY

The 'how often' and 'when' of interest calculations have an effect on the cost of a loan or the value of an investment. Interest may be calculated annually, monthly or even daily, depending upon the type of account. Mortgages and loans are typically calculated monthly, with the interest added to the amount owing at the end of the month. Fortunately, we don't need to be able to work these things out for ourselves – and you will see why shortly – but let's take a look at an example just to see how the interest period can make a difference.

If you invested £100 at 6%, with interest paid once a year, then at the end of the year you would have £106. Yes, yes, we know it's obvious but hang on a moment.

If the interest was paid monthly, things are a bit different. For a start, you won't get 6% – that was the annual rate. Let's divide it by 12 and call it 0.5% a month. That gives you 50p at the end of the first month, which is added to the capital, so that at the end of the second month you get 50.25p. Not a lot extra, but it builds up. In fact, by the end of the year you would have £106.17. That's a whole 17p extra. OK, still not a lot, but if you'd invested £100,000 at the start, the monthly interest would have accumulated into £170, which will fill your tank a few times. Not surprisingly, the difference is even more marked when you are borrowing money, but help is at hand...

APR AND AER

APR and AER are why you don't need to worry too much about how exactly interest is calculated – these measures give standardized rates.

- APR stands for Annual Percentage Rate and is used with mortgages, loans and credit cards. It tells you the total cost of the loan, including set-up fees and other charges as well as the interest itself, expressed as an annual rate.
- AER stands for Annual Equivalent Rate. Banks use this with savings accounts, to describe the interest they earn. (It's sometimes called EAR – Effective Annual Rate.)

When borrowing money, you must always look at the APR – not the headline interest rate number, because the difference between the two can be very marked. There is often an arrangement fee for setting up the loan – typically around £1,000 or more with a mortgage – which will be added to the debt, and therefore incur interest. Some of the less scrupulous operators – of which there are too many – will charge their headline interest rate on the whole amount, throughout the loan period, rather than charging interest only on the amount still outstanding. A certain online loan company, for example, is currently offering loans from '7.5%', though if you read down to the bottom of the page, you will see that the APR can vary from 7.5% to 175% and the typical APR is over 40%. (If you don't meet the repayments, a £5,000 loan at this rate will double in less than 2 years, and reach over £50,000 within 7 years!)

Base rate

The base rate is the interest rate at which the Bank of England will lend money (under certain conditions) to banks. It underpins the rate at which banks will lend money to you or me, and also the amount of interest that banks will pay on the money you lend them. As a general rule, the interest rates that the banks pay on savings are close to base rate; the rates they charge on loans are a few per cent over. Mortgages are typically 3 to 6% over base rate; personal loans 5 to 9% over. There seems to be no visible connection between bank base rate and what credit card companies charge.

Savings

You saw earlier how a wedge of cash can double – perhaps sooner than you expected – simply by leaving it in the bank. A regular savings plan will build capital faster, and we all need a little nest egg (or better still, a large one). With a regular savings plan, the capital builds from the fixed or variable amount you pay in each month, and the interest that is added to it, monthly or annually. In the first few years, the interest build-up isn't that impressive, but over time there is a steady increase in the proportion of the capital that comes from interest.

ISAs – INDIVIDUAL SAVINGS ACCOUNTS

To encourage people to save, the government introduced ISAs back in 1999. These are tax free – you do not pay tax on the interest, and that can make a real difference to the value of your savings. On an ordinary savings account offering 3% interest, you would get to keep £2.40 per £100 as a basic rate taxpayer; a higher rate taxpayer would get to keep only £1.80.

ISAs are aimed at small investors. There is a limit to how much you can invest each year – currently £5,100 – but this can build nicely over time. Here's what happens if you invest the maximum of £425 a month, at the typical current rate of 3% – and look how much difference the tax free status makes:

Years	*ISA capital (interest)*	*Ordinary savings account (interest)*
5	£27,500 (£2,000)	£26,700 (£1,700)
10	£59,400 (£8,400)	£55,800 (£4,800)
20	£139,500 (£37,500)	£122,600 (£20,600)

We've been talking about Cash ISAs, but there is a second type where the investment is in stocks and shares. The interest and capital gains on these are also tax free, but as we all now know, any form of investment in the stock markets carries no guarantee – your nest egg can shrink, instead of growing.

Practice 1

You'll need an online saving calculator for these. There's a good one at the Halifax's site: http://www.halifax.co.uk/savings/calculator. You'll find a link to this and several others at the Crap At Numbers website (www.crapatnumbers.net).

1. Mac's moped is getting near the end of its life. He reckons the string will hold it together for another three years at best. If he saves £100 a month for the next three years, at 2.5% interest, how much will he have to spend on a new one?
2. Heidi is saving for the deposit on a house. She can afford to save £500 a month. If she puts it in an ISA at 3%, how long will it take her to reach £30,000?

Credit cards

A credit card is a convenient way to pay, whether it's online or over the counter, and it's safe as there is a built-in guarantee if the retailer fails to deliver. But using a card can be expensive. Credit card companies make their money from you in four ways:

- A commission (typically 2%) paid by the retailer when you buy with the card, but the retailer will push the cost back to you one way or another.
- The interest on what you owe the company. The basic rule is that when you buy something it does not incur interest until after you've had a chance to pay for it and that can mean up to two months' free credit. (Buy near the start of the month; it won't be added to your bill until the end of the month, and you will have until the end of the next month to pay it.) However, if you have any monies owing at the start of the period, some companies will begin to charge interest on new purchases as soon as they occur.
- Default charges for late payment.
- Charges for cash withdrawals from ATMs, for handling cheques, for foreign currency transactions and similar services – much of which you would get free or cheaper through a current account.

The question is, how can you make money from them, or at least limit the amount they make from you?

PAY UP!

Because you get at least a month without interest on your purchases, buying on a credit card is actually a good way to manage your spending, but to get the benefit, you must pay off the balance each month in full and on time, so that you do not incur interest charges or late payment charges. Ideally, you should set up a direct debit to clear the balance automatically a couple of days before the due date.

If you cannot afford to pay the whole lot off immediately, try to clear the debt as quickly as possible because their interest rates are high. At the time of writing, the bank base rate is 0.5% – but that's what banks charge each other. A bank loan will cost around 10%, which is not cheap, but it's a lot less than the typically 17% APR you pay on credit card balances.

BALANCE TRANSFERS

If you have built up a big balance, or have used the card to spread the cost of a large purchase, do a balance transfer. Many companies offer 0% balance transfers to tempt you to use their cards. Typically they will charge no interest on the transferred balance for 12 or 15 months – sometimes longer – after which they will charge, and sometimes at a higher rate than new spending. (Though note that many do charge a transfer fee of 2% to 3% on the balance, but this is a one-off charge.) This sounds like a good deal, and it is *as long as you do not use the card for any new spending*. Why? Because they normally charge at quite a high rate for new purchases. And you also need to make sure you pay at least the agreed minimum every month, or again penalty clauses can kick in.

Credit card companies can afford to offer 0% loans – which is what this deal gives you, in effect – because they know that many people will not be able to resist using their new card.

HAGGLE OFF THE CARD

This is worth trying. When you are about to make payment, ask if a credit card is acceptable. If it is, offer to pay by cash or debit card instead and ask for a 2% discount. For debit card transactions – as long as they are done online through chip and PIN – the retailer is charged only a few pence. Cash incurs no handling charges. You won't get far haggling over a Mars bar, but it might be worth it for large items of furniture and the like.

CASHBACKS

Some companies offer cashback incentives for purchasing with their cards – typically 0.5% to 1%. So, if you spend, say £1,000, through the card, you would 'earn' up to £10. Put the bulk of your spending through the card and it could build up to a nice little chunk. And why do they give you cash back? Because they hope that you won't pay up on time, and then they will start earning lots of lovely interest from you.

WHICH CARD?

That's not for us to say. The offers change constantly, so you need to shop around for good value. But there are websites that will scour the market for you and offer advice. These are well worth visiting.

- www.moneysavingexpert.com is full of good advice. Use it to work out what you need.
- www.moneysupermarket.com, www.gocompare.com and www.ComparetheMarket.com all compare the available cards in a variety of ways to help you identify the best value.

Do note that most financial advice and comparison websites, though independent, are profit-making ventures and make their money from commissions on sales that come through them, or from advertising or both. And we're not making anything out of any visits you make after reading this...

Mortgages

A mortgage is the second-most expensive commitment most of us will ever take on – having children is the first – so you do need to get it right. Mistakes can be expensive.

The basic theory behind mortgages is simple. You borrow a huge amount of money from the bank or building society, pay it off slowly over a very long time, so that at the end you have a home which you can live in rent-free. If you don't keep up the payments, the bank/building society takes it from you, sells it, takes from this what you owe it and gives you what's left – if anything.

But there's another theory behind the mortgage thing: house prices will keep on rising, so the sooner you get onto the housing ladder, the better. You buy your first home – as a singleton or young couple – then when you are ready to start a family, or have acquired so much stuff the house is full, you sell that house and use the profit from its increased value as a deposit on the next house. You might do this a couple of times until you've got a very nice place. Then – and you're pushing 50 by this time – you remortgage the house to free up the profit from its rising value and buy a little place in the sun. That theory works very nicely as long as (a) interest rates are low, (b) lending money is readily available, and (c) everybody believes in it. If (a) or (b) turn sour, then prices start to fall, (c) is no longer true and prices will drop further. This can leave borrowers with a house that is worth less than the mortgage, which they may well be struggling to repay, and if they sell it they will still be in debt.

This has happened three times in the past 50 years, and the theory is out of favour at the moment. But people forget, and we'll soon be back on the cycle...

The all important LTV

LTV stands for Loan To Value – how much are you borrowing in relation to the value of the property. For the

(Contd)

lenders, this is about risk. They need to know that if you default on the loan they can sell the property and get their money back. So, the bigger the deposit you can put down, the smaller the LTV and the happier they are to lend you money. Right now, if you need to borrow over 75% of the value, you can expect to pay an extra 2% or 3% on the interest rate. This makes it very hard for first time buyers to get into the market – if you can't find maybe £40,000 or £50,000 as a deposit, you will have a very expensive monthly repayment.

There are four main types of mortgages. Let's have a quick look at them.

TRACKER AND STANDARD VARIABLE RATE MORTGAGES

With these, the interest rate varies over time. With a tracker mortgage, the interest is directly linked to the Bank of England base rate – typically 2% to 5% above. With a Standard Variable Rate (SVR) mortgage, the link is not so direct, and any falls in the base rate may take a little while to be followed – though rises tend to be picked up quickly.

Tracker and SVR mortgages tend to be cheaper than fixed rate ones (see below), but the monthly repayments may well change, and if they rise significantly and you are on a tight budget, this can be a real problem. If you have some leeway in your budget, and it looks like interest rates are likely to stay at the current level or fall in the next couple of years, then it's worth looking closely at these.

Discounts and caps

Many companies offer discounted (by around 2%) or capped (with a fixed top limit) tracker or SVR mortgages, as an incentive to new customers. There's a catch, of course. The discount or capping is only for a fixed term – typically 2 or 3 years – and then the normal rate will kick in, and you are committed to staying with the company for another 2 or 3 years after that. You can get out of these deals, but there is always a hefty early repayment charge.

Check the APR

If a mortgage has an introductory offer, the lender must also show an APR figure to cover the whole period. Obviously, with tracker and SVR mortgages, the APR is only a guide figure as the interest rate will vary over time, but it will give you a basis for comparing trackers and SVR mortgages with each other.

FIXED RATE MORTGAGES

With these the interest rate, and therefore the monthly repayments, are fixed for a set period of time – typically 2 or 3 years – after which the mortgage will become a standard variable rate one. At the time of writing, when interest rates are at an historic low, they are a little more expensive than variable mortgages as the lenders are covering themselves against rises in rates. (Remember that they have to borrow the money they lend to you.) Currently fixed rates are about 0.5% higher than the standard variable rate for a 2 year deal, or 1.5% for a 5 year fixed rate.

The big advantage is that you know exactly how much your repayments are going to be over the period of the deal, and if you are working to a tight budget, that matters. But watch out for extended tie-ins. With some, you are committed to staying with the lender on the variable rate for some time after the end of the fixed rate period, with early repayment fees if you want to switch to another lender.

Discounted fixed rate deals are available, but you will find they lead into more expensive variable rates, and that the exit fees are higher.

Arrangement fees

Most lenders charge a fee for setting up the mortgage, and most of them charge more if the mortgage is discounted, or capped, or some other special offer. Watch out for these – and remember that the cost of the fee will be reflected in the APR.

INTEREST-ONLY

With a normal fixed or variable mortgage, your monthly payment covers the interest and a little bit of the capital. (In the first few years the amount of capital you repay is dishearteningly small, but stick with it as it does get paid off increasingly fast.) An interest-only mortgage does just what it says – pays only the interest and leaves the capital untouched. These were popular during the last house price boom because they are cheap and people could expect to sell them after a few years and get a nice chunk of capital from the rise in prices. When prices are static or falling, they lose that attraction. Unfortunately, a lot of people took out these during that last boom, and are now having to live with the consequences.

There has to be some way of paying off the capital, so if you want to own the property eventually, you have to set up a saving scheme to run alongside, and your total outgoings will be the same or more than with a normal mortgage.

Interest-only mortgages are now used mainly by buy-to-let investors, who can claim tax relief on the interest payments, while they draw an income and/or build capital from rents.

OFFSET MORTGAGES

Offset mortgages are for people with savings. The deal is this. Instead of having savings in a separate account, earning a bit of interest, they go into an account linked to the mortgage to reduce the amount you owe. A current account can also be linked in the same way. The benefit – and it is a substantial one – comes from swapping your savings interest (which would be subject to tax) for a reduced interest on the loan (which you are repaying out of after-tax income). You could get the same benefit, of course, by using your savings to reduce the mortgage at the start, but this way you still have access to your savings, if they are needed.

If you are a higher rate tax payer, have a good chunk of savings, and tend to run with a healthy balance in your current account, this can be a very attractive option. Here's a much simplified example.

If you had £10,000 savings, then at the current rates of interest and tax, this would earn about £6,500 over 25 years, but you would pay nearly £2,500 tax – an overall gain of around £4,000. Paid into an offset, with a mortgage rate of 3.5%, it would reduce the total cost by over £10,000.

Shared ownership

Shared ownership offers a possible way into the housing market for those many first time buyers who don't have the substantial deposits that are demanded nowadays. The schemes are run by some housing associations, and a few of the larger builders. The resident buys a fraction (typically 25%) of the house, and pays rent to the builder or housing association to cover the rest. When the house is sold, the resident takes the same fraction of the proceeds, and in a rising market this could be enough for a deposit on the next house.

Note that some shared ownership schemes are only open to certain occupations.

CHOOSING A MORTGAGE

We are not financial advisers, so we can't help you make the choice beyond what we've already said. There are lots of people out there, online, on the phone and in the high street, all anxious to give you advice, but do remember that most of them are doing this for a living.

- The banks and building societies will help you to choose the mortgage that best suits your situation – but only from within their own range.
- A mortgage broker will draw from the wider field, and FSA (Financial Services Authority) registered brokers are quite tightly regulated to ensure impartiality and fairness to their clients.
- Online comparison sites will show you what's around across the whole field, and normally have online calculators that will give you a good idea of actual costs. Some of these sites will pass your requirements on to lenders and collect quotes for you, sending them by email.

USEFUL WEBSITES

- www.direct.gov.uk – Completely impartial advice from the government. **Head for Money, tax and benefits, then Managing money**, then **Mortgages.**
- www.moneymadeclear.org.uk is run by the Consumer Financial Education Body (government funded). It promises 'No selling. No jargon. Just the facts.' And delivers just that. It offers product information, impartial advice, comparison charts and calculators.
- www.mortgagesorter.co.uk was one of the first websites in the online mortgage advice field and has established a good reputation.
- www.moneysupermarket.com, www.beatthatquote.com, www.gocompare.com and www.ComparetheMarket.com all offer similar services, with comparison charts and calculators, and quotes available via email.

Buying and leasing a car

People buy cars for a number of reasons. These may include any or all of the following:

- For the daily commute to work or the school run
- For doing the shopping and other little trips around town
- To take the family on holidays or long distance visits
- As a status symbol
- To carry samples, equipment, musical instruments or other stuff to clients or gigs
- To express their personality
- To cover up their lack of personality.

The reasons for buying a car affect the choice of vehicle – obviously – but also its age and how to pay for it. At the one extreme, are people who always buy new cars, replacing them every 2 or 3 years, because they need a reliable vehicle or they need to impress clients, or they like the smell of new cars, or whatever. At the other extreme are those who just want something that will get them (and their family and their stuff) from A to B as cheaply as possible.

If it's a new car you want, there are basically four ways of funding it:

- Finance through the car retailer/manufacturer
- With a bank loan
- From your savings
- Lease/purchase.

Finance

Finance through the retailer or manufacturer is often the simplest to arrange – you are there, in the showroom after choosing the car, and the finance can be set up at the same time. The APR on these is typically around 9% at present, though there are special deals around. Firms will sometimes offer 0% finance deals when they are very keen to shift stock.

At the moment, a typical deal looks like this:

Cash price	£12,500
Deposit	£2,500
Balance	£10,000
Arrangement fee	£150
36 monthly payments of £320	£11,500
Total amount payable	£14,150

Bank loan

A normal bank loan will work out at much the same cost as a finance deal – though it's always worth shopping around for the best rates. However, with a prearranged loan in your pocket you may be able to negotiate a discount and that would reduce the amount you would actually need to borrow.

Cash purchase

Buying from your savings will give you a better deal, though you have to have the savings to start with, of course. But look at it in the long term. If you intend to change your car every three years, you will be making repayments all the time – so what happens if you saved regularly at the start, and got ahead of the game?

At an interest rate of 3% you would only need to save £270 a month to build £10,000 in three years. That's £50 a month less than the loan repayment.

The next car

All these calculations are based on the initial purchase of a £12,500 car with a £2,500 deposit. If that car is traded in for a new model in three years' time, its trade-in value will be considerably more than £2,500. Some cars depreciate faster than others (no matter who drives them), so the trade-in value could be anywhere between £5,000 to £8,000. Buying a similar level of car next time will cost you less.

Lease/purchase

Many car firms offer lease/purchase financing. With these deals, your regular payment is considerably less than with a standard finance arrangement, but that is because you are not actually buying it. You are in fact leasing the car, with an option to buy it at the end of the period.

Heidi is looking at a Smart car. Smart has a lease purchase scheme, and the figures currently look like this for a Cabriolet:

Cash price	£10,600
Deposit	£1750
Monthly costs: 36 × £130	£4680
Option purchase price	£5,500

If you choose to buy your car at the end of the period, you will have paid £12,000 in total. If you don't take up the option, you will have paid £6430, but will then need to start another lease deal if you want to stay on the road.

Lease/purchase is mainly used by firms for keeping their company car pool stocked, but it is an option to consider, especially if your circumstances may change in the next few years.

Practice 2

1 Mac has decided it's time to move on from his moped and wants a Harley Davidson Electra Glide (in blue). It will cost £25,000 and he's managed to save £5,000. The advance on this book won't pay for it – in fact, it won't even fill the petrol tank, so the question is: how to pay for it? Use the Loan Calculator at the Motley Fool (http://www.fool.co.uk/loans/loan-calculator.aspx) to find out how long it will take him to pay for the bike at £250 a month, when the interest rate is 8%.

Calculating tax

Benjamin Franklin once famously wrote, 'In this world nothing can be said to be certain, except death and taxes.' Of course, he hadn't heard of non-domicile arrangements and they didn't have offshore tax havens in 1817. However, taxes (and we're talking about those on income here) are inevitable and unavoidable for us little people, so the only real question is, 'How much will they be?'

The first point to note is that there are two taxes on income: the one called income tax and the one called National Insurance contributions. They are calculated in different ways – and the numbers change every year, so we are not even going to try to show you how to calculate your tax exactly. What we will do is show how it works and give you a way to work out a rough figure. You can get the exact up-to-date figures easily enough from one of the many on-line tax calculators listed at the end of this section.

INCOME TAX

This is a progressive tax – in theory, the more you earn, the more you pay – and it's based on annual income, though you may be paid monthly or weekly.

At the time of writing, here in the UK, approximately the first £6,500 of income is free of tax. Any income up to £37,400 above that is taxed at the basic rate of 20%; income above that, up to nearly £150,000 is taxed at the higher rate of 40%, and anything over that is taxed at 50%. (We've rounded the figures to make them easier to work with, and with another budget in the offing as this book goes to press, they will have changed anyway. The important thing is the process – substitute the correct up-to-date figures as needed.)

Let's see how that translates in real terms (and when we say 'real' we mean 'fictional'). When we are not writing books, Mac is a bagpipe repair man on £17,500 p.a., and Heidi works for a Swiss bank, on £75,000. We'll work up from the bottom.

MAC'S INCOME TAX CALCULATION

	Annual	*Monthly*
Income	17,500	17,500/12 = 1,460
Tax free	6,500	
Basic rate band	17,500 – 6,500 = 10,500	
Basic rate tax	10,500 × 20% = 2,100	
Higher rate income	0	
Higher rate tax	0	
Total tax	2,100	2,100/12 = 175
Remainder	17,500 – 2,100 = 15,400	15,400/12 = £1,280

HEIDI'S INCOME TAX CALCULATION

	Annual	*Monthly*
Income	75,000	62,500
Tax free	6,500	
Basic rate band	75,000 – 6,500 = 68,500	
Basic rate tax	37,400 × 20% = 7,480	
Higher rate income	68,500 – 37,400 = 31,100	
Higher rate tax	31,100 × 40% = 12,440	
Total tax	7,480 + 12,440 = 19,920	19,920/12 = £1,660
Remainder	75,000 – 19,920 = 55,080	55,080/12 = £4,590

Heidi's just checked her last payslip, and her tax was far lower than that. What was wrong with our calculations? Nothing – if she had worked the whole year at that rate. The tax year runs from the start of April in one year to the end of March in the next year, and allowances and bands are for income during that period. As it happens, Heidi didn't start work until September, having spent the first 6 months of the year in Tibet, so by March, she'd only earned a total of £37,500 but she has a whole year's income tax allowance to spread over those last 6 months. Let's have another look – and this time, you'll see why we've had a monthly column in these tables.

HEIDI'S INCOME TAX CALCULATION (MARK II)

	Annual	*Monthly*
Income	37,500	6,250
Tax Free	6,500	
Basic rate band	37,500 – 6,500 = 31,000	
Basic rate tax	31,000 × 20% = 6,200	
Higher rate income	0	
Higher rate tax	0	
Total tax	6,200	6,200/6 = £1,030
Remainder		6,250 – 1,030 = £5,220

Next year will be different, as the higher rate will be payable, and will kick in straight away.

The first payslip

If you are employed (and not self-employed), you pay your tax by PAYE (Pay As You Earn). This works on a cumulative basis. A tax free allowance of £6,500 p.a. is about £125 a week or £540 a month. The tax year runs from the start of April in one year to the end of March in the following year. You can earn £540 tax free until the end of April, £1,080 to the end of May, £1,620 to the end of June, and so on. If you are out of work, or out of the country, or otherwise not earning at any point during the year, your tax free allowance builds up while you are away, and is taken into account when you start work again. The amount that's due for tax at basic rate likewise builds up month by month. £37,400 p.a. is about £3,100 a month, so you can earn £6,200 to the end of May and still only be paying tax at 20%.

The effect of this is to give a nice little boost to the first paycheck. (Though it sometimes takes until the second or third for your employer to get your tax position sorted out properly, and in the meantime you will be taxed as if you'd been earning all year.) In Heidi's case, by the end of September she had built up 6 months' worth of tax free allowance – that's £3,250, so her first paycheck works out like this:

HEIDI'S FIRST PAYSLIP

	September
Income	6,250
Tax free to date	3,250
Taxable income	3,000
Basic rate allowance	3,100 × 6 = 18,600
Basic rate tax	3,000 × 20% = 600
Higher rate income	0
Higher rate tax	0
Total tax	600
Remainder	6,250 – 600 = 5,650

But don't forget the government takes another slice...

NATIONAL INSURANCE

Paying National Insurance Contributions (NICs) entitles you to the state pension and benefits when you are out of work or ill. It is not strictly speaking a tax, but everyone in work and under pension age has to pay. It is based on weekly income (if you are employed) or annual income (if self-employed). There are lots of different categories and special conditions, but for most of us normal working stiffs, the deal is:

Below £110 a week	Under £5,720 p.a.	no contribution
£110 to £844 a week	£5,720 to £43,888 p.a.	11% of income
Above £844 a week	Over £43,888 p.a.	11% on earnings up to £844, + 1% on earnings over £844

But note the emphasis on weekly. The fact that Heidi had not worked for half the year did not make any difference to her contributions.

Let's add these into the calculations and see how much we've got left to spend.

Mac's £17,500 p.a. works out at £336 a week, his income tax is £40 a week.

11% of 336 = £37 a week NIC

Take home pay: £336 – 40 – 37 = £259

Heidi's £75,000 p.a. works out at £1440 a week, her tax (last year) was £240 a week.

11% of £844 = £93
1% of (£1440 – 844) = £6
NIC = £93 + 6 = £99

Take home pay: £1,440 – 240 – 99 = £1,101

The rough check

If you are heading into a new job, or have a nice pay hike ahead, there are online calculators (see next) which will give you a reasonably accurate figure for how much tax you'll have to pay. But for a rough idea, quickly, here's the Heidi and Mac 'What's it worth?' rule-of-thumb.

- With jobs under £15,000 the taxman will take under 1/5.

(Contd)

- With jobs paying between £15,000 and £40,000 p.a., tax and NI contributions will take approximately 1/4 of your income – less at the bottom end, a bit more at the top. If you get a £1,000 p.a. raise you will keep about £700 of it.
- With jobs between £40,000 and £150,000 the taxman will take 1/3. A £10,000 p.a. rise puts just under £600 in your pocket.
- With jobs over £150,000 you will employ tax advisers whose fees will be less than the tax they save you.

ONLINE CALCULATORS

If you want to work out your tax, there are calculators at these websites:

- www.incometaxcalculator.org.uk straightforward, fast and simple to use. This will work from weekly, monthly or annual wages for the current year, last year and next.
- www.moneysavingexpert.com/tax-calculator this has some extra frills, e.g. you can work in pension contributions and student loan repayments, so may produce a more accurate result.
- www.listentotaxman.com another no-frills calculator, but this one is available as an iPhone app.

The life sum: when can I retire?

The government says we should all make provision for our old age. Well I've made mine. I've decided I'm going to be a burden on people.

Andy Hamilton

You can, of course, retire whenever you like. The question is: 'When can I retire if I want to eat and keep warm?' or 'How can I secure a good pension?'

Apart from the state old age pension, that we all get if we've paid our NI contributions, but won't feed us well or keep us warm through the winters, there are two types of pensions:

- Final salary schemes
- Private pensions, which may be personal or company

Both types are based on the principle of accumulating a pot of money during your working life, which is used to buy an annuity. This is in effect the reverse of a mortgage, as it turns capital into income. How much income you get for your pot depends on two things: the interest rates at the time and for the foreseeable future – which likewise determines the cost of a mortgage – and the likely duration. A mortgage is for a fixed number of years; an annuity continues as long as you are alive. There's a gamble going on here. Whoever is paying the annuity is betting that you'll die sooner rather than later.

The taxman's bonus

And for once, the bonus goes to you, rather than the taxman. You do not have to pay tax on the money that you invest in a pension. If you are in employment, tax is calculated on your income after your pension payments have been taken out. If you are contributing to a personal pension plan, then the taxman gives you tax relief on the payments, boosting the value of the contributions by 25% for basic rate taxpayers, or 50% for higher rate payers.

Of course, he'll get you at the other end. Pensions are subject to tax, though as your income then will probably be less than now and as you will have the personal allowance, and age allowance, etc., the amount of tax that you pay on it will be less. At worst you've postponed the tax into the far future.

FINAL SALARY SCHEMES

These have become increasingly rare in the private sector, and may well be phased out of public sector jobs in the future, as they can be expensive to run. The basis is that for every year of service in the job, you receive 1/80th (or 1/60th in some jobs) of your final salary as a pension. A teacher, for example, starting work at 23, and retiring at 63 as a deputy head on, say £50,000 p.a., would have a pension of 40/80 × £50,000 = £25,000. Teacher's pensions – like most others – also pay out a lump sum of 3/80th for every year's service, so our retiring deputy would have £75,000.

Most final salary schemes are index-linked. If the link is to salary, the pension will go up in line with the salary paid to deputy heads. If the link is to prices, it will go up in line with the Retail Prices Index, or some other agreed measure.

These are the best sort of pension schemes, but they are not free, of course. Throughout your working life, you will be paying into them – typically around 6% of salary. Of itself, this would not be enough to fund the pension but the employer also chips in 6% or more. Even then, the contributions may still not be enough, as people are living longer. For example, up until around 20 years ago, the teacher's pension fund ran at a profit to the government, simply because the average teacher died not that long after retiring – 40 years at the chalkface takes it out of people. Earlier retirement, better work-to-retirement planning and general better health mean that a lot more ex-teachers now draw their pensions a lot longer, and what they paid in during their working life doesn't provide a big enough pot to cover the pension. With public sector jobs, the taxpayer picks up the shortfall. With private sector final salary schemes there isn't that fallback. We'll have a look at private pensions in general then get back to this.

Index-linking and inflation

Inflation has been a fact of life since the invention of money. All that has changed is the rate. Inflation erodes the value of savings and of fixed pensions. At 3% it will halve the value in 25 years; at 5% it will halve in 16 years. A person retiring at 60 is very likely to live for another 20 years or more, which is why index-linking is such a good idea.

The link can be to the following:

- The Consumer Price Index (CPI) is a measure of inflation calculated from the average price increase of a basket of 600 different goods and services.
- The Retail Price Index (RPI) is based on a similar basket to that used for the CPI, but also includes housing costs. In recent years the RPI figure has tended to be higher than the CPI figure – reflecting the big hike in house prices and rents. There is also an RPIX figure which excludes mortgage interest.
- Salaries, which tend to rise faster than inflation.

PRIVATE PENSIONS

Company and personal pension schemes are all based on investment. The contributions go into a fund which is invested in the stock market and money markets. When the policy holder reaches retirement, the value that has built-up in the fund is used to purchase an annuity to pay the pension. During boom years, rising share prices multiply the value of the pension pot, but – as it says in the small print, and as many people have discovered to their cost – share prices can go down as well as up. In a falling stock market, individual policy holders find that when they come to retire, the pension is far smaller than they had been led to expect during their paying-in years, and companies find that they do not have enough in the funds to cover the promised pensions, and so must reach into their profits to pay the difference. It's the long-term cost of under-funded pensions that is pushing companies to abandon final salary pension schemes (though they seem to keep them going for directors).

What will a decent pension cost?

How much? That depends upon how much you think a decent pension should be, at what age you intend to retire and whether you are male or female. It will also vary with current and forecasted interest rates. (And it's worth adding that when it comes to the crunch, your postcode, health and other factors which can affect your longevity will all be taken into account.)

(Contd)

At the time of writing, here's how much of an (index-linked) annual pension you would get for £100,000 – multiply up to suit your retirement lifestyle.

Retirement age	*Male*	*Female*
60	£3,300	£3,000
65	£4,000	£3,600
70	£5,000	£4,500
75	£6,500	£5,600

Answers and explanations

Practice 1

You'll need an online saving calculator for these. There's a good one at the Halifax's site: http://www.halifax.co.uk/savings/calculator.

1 In three years, £100 a month at 2.5% interest will produce £3,734.40.

2 It will take Heidi 4 years 8 months to save £30,000 at £500 and 3%. Unfortunately, you can't put that much a year into a cash ISA – the limit is £5,100 a year, or £425 a month.

Practice 2

1 The calculator tells us that £20,000 over 9 years at 8% would cost £257 a month; over 10 years it would be £240 a month. So, paying £250 a month should clear it in about 9.5 years.

Taking it further: living numeracy

USEFUL ORGANIZATIONS

Numeracy qualifications and courses

Learn Direct: an e-teaching organization set up by the University for Industry in 1998, and a good place to find out more about courses and qualifications. Contact points:

- Online at: http://www.learndirect.co.uk
- Telephone 0800 101 901 (England and Wales) or 0808 100 9000 (Scotland)
- Post learndirect, PO Box 900, Leicester, LE1 6XJ

The **BBC** offers online learning and support for adults in maths (and much else besides). Go to: http://www.bbc.co.uk/learning/subjects/maths.shtml

The BBC Bitesize materials are designed for GCSE revision, but can be a good way of mugging up on any maths topic at that level. Head for: http://www.bbc.co.uk/schools/gcsebitesize/maths/

Edexcel has qualifications in adult literacy and adult numeracy (ALAN). You can find out more at: http://www.edexcel.com/quals/skillsforlife/alan/Pages/default.aspx

The Open University runs an excellent Maths Help website as part of its OpenLearn programme. It's free, nicely presented and very accessible. Go see at http://mathshelp.open.ac.uk/.

FURTHER READING

Basic Accounting, Andy Lymer and Nishat Azmat (Teach Yourself) 2010

Basic Mathematics, Alan Graham (Teach Yourself) 2010

Get Started in Computing, Moira Stephen (Teach Yourself) 2010

Get Started with Excel 2010, Moira Stephen (Teach Yourself) 2010

Improve Your Cash Flow, Alan Warner and Robert McCallion (Teach Yourself) 2010

Plan A Secure Retirement, Trevor Goodbun (Teach Yourself) 2010

Sort Out Your Family Finances, Bob Reeves (Teach Yourself) 2010

Thrifty Living, Barty Phillips (Teach Yourself) 2010

Understand Algebra, Paul Abbott (Teach Yourself) 2010

Understand Calculus, Paul Abbott and Hugh Neill (Teach Yourself) 2010

Understand Statistics, Alan Graham (Teach Yourself) 2010

Understand Trigonometry, Paul Abbott and Hugh Neill (Teach Yourself) 2010

Index